Venice and the Water

For Joyce & Jan
the ecological history
of a magical place!
— Chuck

Venice and the Water

A Model for Our Planet

Piero Bevilacqua

Translated by

Charles A. Ferguson

With an Afterword by
Massimo Cacciari

Polar Bear & Company
Solon, Maine

Polar Bear & Company
P.O. Box 311, Solon, Maine 04979 U.S.A.
207.643.2795 www.polarbearandco.com

First English language edition 2009
13 12 11 10 09   1 2 3 4 5 6 7
Cover art & design, Emily Cornell du Houx; Maps created by Charles A. Ferguson and Paul Cornell du Houx, based on maps published by Istituto Geografico DeAgostini, Touring Club Italiano, and Google.com.
This selection was originally published in Italy by Donzelli Editore under the title *Venezia e le acque*, copyright © 1995, 1998, 2000 Donzelli Editore.

Library of Congress Cataloging-in-Publication Data

Bevilacqua, Piero, 1944-
[Venezia e le acque. English]
Venice and the water : a model for our planet / Piero Bevilacqua ; translated by Charles A. Ferguson ; with an afterword by Massimo Cacciari. -- 1st English language ed. 2009.
p. cm.
Includes bibliographical references.
Summary: "A history of political checks and balances that developed in the Italian city of Venice in dealing with the environmental challenges of the city and the lagoon. The priorities of a scientific approach combined with a sense of the common good are held as a model for rescuing the planet"--Provided by publisher.
ISBN-13: 978-1-882190-59-1 (pbk. : alk. paper)
ISBN-10: 1-882190-59-9 (pbk. : alk. paper)
1. Venice (Italy)--Threat of destruction. 2. Water levels--Italy--Venice, Lagoon of. 3. City planning--Italy--Venice. 4. Cultural property--Protection--Italy--Venice. I. Title.
DG672.5.B4813 2009
363.7--dc22
2008049459

Manufactured in the U.S.A. by Thomson-Shore, Inc., an employee-owned company—certified by the Forest Stewardship Council and a member of the Green Press Initiative—using soy ink and acid-free, recycled paper of archival quality, at paper permanence specifications defined by ANSI.NISO standard Z39.48-992: "The ability of paper to last several hundred years without significant deterioration under normal use and storage conditions in libraries and archives."

For my mother, Agata

In the makeup of the globe, water is the great demolishing and spreading force: by erosion it breaks down and flattens, and by deposit it fills and evens. The enemy of loftiness, elevation, and inequality, it is the democratic element par excellence, working to make the world smooth and round.

—F. Porena, *Sul deperimento fisico della Regione italica*, 1896

CONTENTS

Translator's Note

The earliest settlements in the Venetian Lagoon date back to the 4th century C.E. On a cluster of tiny islands called "Rialto" (*rivo alto*, high bank), a trading post took root and eventually, with the name "Venezia," grew into a powerful city-state. By 1300, Venice controlled Europe's trade with the Near East and Asia. (The most famous commercial traveler of all time, Marco Polo, was a Venetian.) Venice was one of the largest cities in Europe, and she was unique in that she had no walls for her defense, as her government liked to boast, just "salt water." Her Arsenale (shipyard) was the largest industrial complex in the Western world. Today tourists from all over the world flock to Venice because of her unique beauty, just as in past ages she attracted pilgrims and scholars, traders and diplomats, because of her prestige, her wealth, and her power. The City in the Lagoon was unique with respect to her government as well. Throughout her long, independent life, Venice was the only state in Europe that was not a monarchy. La Serenissima (the Most Serene Republic) stood proud and sovereign for 1300 years, until Napoleon set up a Kingdom of Italy and in 1797 abolished Venice's independent regime.

As the title suggests, *Venezia e le acque* studies the role of ecology in the long history of Venice. The author argues that Venice's rise to greatness despite a difficult environment was another unique achievement of the City in the Lagoon, and perhaps her greatest. Since 1797, however, Venice has declined, and her environment has been drastically compromised in the interest of progress. In his preface Bevilacqua devotes a fair amount of space to a critique of Western society today, charging that consumerism and a distorted idea of progress have discredited the study of economic geography and the history of technology. That indifference, he contends, influences teaching and scholarship, leaving Italians shockingly ignorant about the achievements of their own country. A further distorting influence is what Bevilacqua describes as the humanist's prejudice against technical and practical matters. He sketches a wry picture of Italian political life, where technical competence is trumped by empty rhetoric, and in this context he sees the prospects for restoring Venice's environment as dim. I have retained these polemical pages partly because they give one Italian's perspective on traits we may think of as "typically Italian," but most of all because they have their relevance to public and academic life in the United States. It is left to Massimo Cacciari, a philosopher and recently the energetic mayor of Venice, to provide an afterword of hope and resolve.

Bevilacqua shows that throughout the centuries of her sovereignty Venice worked to reconcile private interests with those of a state for which—here more than anywhere else in Europe—preserving the environment was a matter of life and death. The world that sustains our lives today may seem reliable, infinitely more durable than a cluster of islets surrounded by marshes and a tidal lagoon. Nevertheless, we know that our planet is drifting closer to ecological disaster every day. La Serenissima left us a model for reconciling human enterprise with the environment, a legacy of public policy that for centuries sustained a city's enterprise and her very life. I hope that by making Bevilacqua's essay available in English I may help to bring our world a step closer to rescue.

Note: In documenting this vast survey, the author cites a wealth of sources, both ancient and modern. Although with few exceptions his bibliography is in Italian, we have included Bevilacqua's notes because they give additional glimpses of Venetian life over the centuries and perspective on the development of today's ecological crisis.

Charles A. Ferguson
Associate Professor Emeritus
Colby College

Preface to the Third Edition

The second edition of this book, published in expanded and revised form in 1998, benefited from an unexpected advantage: a timeliness that could not have been foreseen. The book came out just when the debate over how to save Venice was growing more intense in Italy—with far-reaching echoes internationally—specifically, whether to adopt or reject the project called Mo.S.E (Experimental Electro-mechanical Module), with movable mechanical barriers at the seaward passages *[bocche di porto]* to protect the city and the lagoon from flooding by the Adriatic. On 8 July of that year, an International College of Experts submitted to the prime minister and to the ministers for the environment and public works its report favoring the project, bringing to the wide-ranging debate further topics for consideration and study. Our book did not take a position among the opposing groups—though expressing hope that the Mo.S.E. project might succeed—and did not set forth options of a technical nature. The dilemmas inherent in the choices to be made and his lack of competence in the matter have increased the author's inclination towards uncertainty; he abstains from a final judgment that should be left to those who can speak about things they know and understand. On the other hand, the timeliness of the historian's work—and the substance of his political message—consist in this case of showing the Venetians' *capacity for decision* throughout all the centuries of the Republic's life. The special, courageous ability of the city's magistrates to make choices, sometimes drastic ones, enabled them to resolve problems that could not be put off. This ability was definitely not born of crude "executivism"; on the contrary, it matured amid long and stormy controversies, debates, assessments, reports, and onsite observations. As the Venetians under the old regime liked to say, decisions to take action and modify the existing balances were made "experimentally," keeping under systematic observation the effects resulting from the changes made, especially in the area of hydrology. Paradoxically, despite the scientific fashion that dominates our fields of knowledge today, a willingness to experiment has disappeared from the thinking and operating horizon of politics and government.

Today, this third edition of *Venice and the Water*—which is unchanged from the second—comes out at a time when the debate over the destiny of the city seems to have dozed off. The public's interest in Venice seems to come and go, like the high tides that flood her more and more frequently. This book—which is sold out—is definitely not being reissued with the aim of stimulating action or reopening the debates over the city in the lagoon. It does, however, bring to the reader its aim, ambitious and explicit, of summarizing once again the exemplary history of how, for so many centuries, the Venetians controlled and safeguarded their environment, even making it the fulcrum of the city's economic prosperity. The lesson is still intact, and it looms over every possible situation in today's developments. Anyone wishing to know how and why the argument goes on, in varying rhythms, over the destiny of one of the most miraculous cities on earth, may find here the basic lines of development leading from past centuries to the present situation.

Rome
March 2000

P.B.

Preface to the 1998 Edition

There are books an author never manages to finish, never frees himself from once and for all. In some cases the events he has undertaken to reconstruct cannot be considered as past; the present keeps them open to unforeseeable developments, and so he must follow them and keep bringing his account up to date. This is certainly the case with *Venice and the Water: A Model for Our Planet*, first published in 1995 by Donzelli in a series called *Saggine [New Brooms]*. The present revised and expanded edition was not published just because the previous edition sold out. Other motives prompted the author to take up the old story again and carry it forward with renewed energy.

The Venetian Republic's farsighted policy for safeguarding the lagoon throughout the modern era was the sole topic of the previous book, which several observers soon judged incomplete. The little book told an amazing story of technical and political achievement on a unique scale: the centuries-old struggle of a city-state to preserve the balance of the special habitat provided by her streams and lagoons. The lagoon enabled Venice to become a maritime center of the first magnitude, although a set of irresistible forces, both natural and social, tended to fill the lagoon and turn it into a malarial marsh, threatening the city with depopulation and abandonment.

To readers, the account of those events and their splendid conclusion seemed more and more incomplete, a cameo too remote from the present in its polished perfection. The tale ended with the fall of the Republic at the end of the 18th century. But seeing the many problems facing the Venetian Lagoon today, the book stirred up a hornets' nest of issues and left them unanswered. What was the sequel to that story? What misfortunes have befallen the lagoon and its largest island during the contemporary period? What forecasts, what expectations, projects, and fears are on the horizon of the near future? Too much history has accumulated during the last two centuries to be left out of the picture. Moreover, the efforts made in recent years to safeguard the lagoon are beginning to write a new chapter in a history that's worth knowing. Most assuredly, Venice's experience is not one that can be bracketed within the barren and conventional periods of academic history.

To answer these questions, I was moved first of all by the legitimate curiosity of a few foreign publishers who had been fascinated by the amphibious experience of the City in the Lagoon throughout the modern era. A French publisher, Liana Levi, urged me to retrace the process leading to today's developments, as a complement or completion to the historical account. So the original essay changed shape and dimensions, becoming an elegant illustrated book with the title *Venise et l'eau*, published in Paris in 1996. Next, Campus Verlag in Frankfurt asked me for a text similar to the French version, which I had expanded in the meantime, for a German edition—illustrated—titled *Venedig und das Wasser: ein Gleichnis für unseren Planeten*. It was issued in the spring of 1998. An Italian version could not fail to appear, taking into account the new material contained in the two foreign editions. Once again, after a few quick visits to the Venice State Archives and another, longer immersion in the contemporary historical literature, a book resulted that was further expanded, retaining all the original materials, of course, but once again differing from all the previous versions by the abundance of its documentation.

Habent sua fata libelli is a maxim that could well apply to the growth of this text, continually evolving and not ending with the publications just mentioned. For a Calabrian to write a history of Venice—even though he bears an ancient Venetian family name—is unusual, especially since historians generally remain close to their native regions. It's even more unusual if, as in the present case, the historian has earned for himself—despite the variety of topics he's treated in twenty years of research—an indelible identity as a student of Southern Italy. Ever since the publication of *Venezia e le acque*, I've had to "justify" to friends and admirers this incursion into foreign territory. What secret curiosity drove me to do research on such unaccustomed shores? And why retrace a chapter of modern history, the sumptuous history of Venice, even from the particular standpoint I've chosen, especially since it has enjoyed the attention of an international crowd of scholars and researchers for decades?

In fact, the project behind this history is not nearly as foreign to the field of my research and interests as it may appear at first glance. Stating it here affords me the opportunity for a few general considerations I should like to include in this new edition intended for Italian readers. The moment when I first glimpsed the idea of studying the Venetian Lagoon is significant for its connection with my basic interest in studying the life of the land in Italy. It came while Manlio Rossi Doria and I were writing *Land Reclamation in Italy from the Eighteenth Century to the Present [Le bonifiche in Italia dal Settecento a oggi]*, published by Laterza in 1984. In the midst of that research, while reading documents on 19th- and 20th-century land reclamation in Veneto, I discovered the historical origins of swamp formation in certain tracts of the mainland. The stream diversions carried out by the Venetian Republic to keep the lagoon from silting up had disrupted the natural hydrographic relationships inland, and the imbalance had never been permanently remedied. Streams that were forced into artificial channels with artificial profiles at the end of their courses tended to flood the countryside periodically. In many cases, swamp formation was the secondary effect of that daring technical operation, carried out by one of the most powerful city-states on the Italian peninsula. So it was that I chanced to discover Venice's lagoon policy, from the evidence of its unwanted effects. But the discovery was no less dazzling. The titanic undertaking of changing the course of some of Italy's major rivers, in order to preserve the balance of a local habitat, was an ambition that could leave no reader indifferent. Nor could it fail to kindle a desire for learning more, in the soul of a non-Venetian historian.

Actually, from my personal standpoint, the story of the lagoon wasn't originally supposed to be presented as a separate account; it didn't have the ambition to take the form of a book. First of all, for reasons of modesty and intellectual caution. Anyone who frequents a scholarly field today knows what a mass of bibliography, monographs, and specialized research he must deal with before he can venture something more than an amateur's judgment. This is almost natural behavior among scholars nowadays, their mental habit. However, some Italian "pirate pens" haven't the faintest idea of research: they're amateurs who've never set foot in an archive, but with brazen naïveté they venture to write revisionist treatments of great historic events, just because they've read a few books about them.

In fact, the lagoon experience was supposed to be one chapter of an imagined *History of Water in Italy*, an old love partly fulfilled by the essay "Water Revolutions," published in the first volume of the *History of Italian Agriculture*, published under my direction by Marsilio in 1989. I've always seen as a huge cultural deficit in our nation its absolute ignorance of our country's hydrography and its evolution. It's a history known only to a few specialists, a few hydraulic engineers, and a few isolated and heretical historians. And yet, Italy owes a great part of her landscape today, the prosperity of many of her economies, and even the essential features of some of her regions, to that evolution. The Po valley has been completely made over by modifications to the chief rivers that flow within it. Canals, springs, aqueducts, channels, etc. have constituted a network of water resources as the foundation upon which one of the most prosperous agricultures in the West has been developed.

In central Italy, the tenant-farm economy has coped for centuries in a different way with the waters of the Apennine torrents: controlling them, channeling them, limiting their erosion, but never giving up the use of them, not only for households and irrigation, but for power to shape the hills, adjusting the fields to slopes compatible with cultivation. Such is the centuries-old work of men and water, giving the Tuscan ridges and hills the slopes and contours that delight us today.

No less exciting was the evolution of streams in the South. For a long time those regions were forced to recognize in the unruly waters a treacherous enemy of all economy and even of human life, but it's no less true that by using hydrographic resources it has been possible to transform agriculture radically. Since the last war, through the construction of great impoundments—for example the huge one built by Agri-Sinni—the South has been equipped with an extensive network of irrigation canals that have increased irrigated land to over 400,000 acres. The historical deficit of Mediterranean agricultures, namely the spring and summer drought, has been overcome and turned into an advantage compared with agriculture in the northern regions. Water brought into the countryside has *reclaimed* the climate of those parts and removed a thousand-year-old natural limitation, multiplying the productive force of the sun's rays on plant biology. Here again we have a gigantic project to which we owe many of the modern and advanced features of river-plain agriculture in today's South. This is another of the historic achievements unknown to our country's culture and public opinion.

We must remember that this cancellation of the past, this erasing of our national memory, is both a new phenomenon and an old inheritance. It's a new and universal fact because the economic demons of our time push individuals, institutions, and cultures to consider the present as forever inadequate, always far removed from the ideal standards imposed as goals to be pursued in a never-ending race. So it is that every goal reached, every progress realized is soon cancelled in view of a new challenge, of further objectives awaiting fulfillment. There is always a region, a country moving ahead of the others that must be imitated. And so the past—the artifacts and the traces of its work, the consolidation of the work of previous generations—is soon demoted to the status of an "outdated reality," in a present that can never stop and observe, can never appreciate itself because it's being gnawed by the demon of competition, busily devouring the future. A mentality of constant devaluation, driven by the essentially destructive character of the world economy today, drives us to see the present as a past that's slow to move on, a universe of value lost.

But at the same time, in Italy, ignoring is an ancient trait, peculiar to a country congenitally unable to recognize its own achievements and the heritage of its own virtues. It's a nation deaf and blind to any appreciation of the achievements of those who went before. Especially if those achievements take the form of artifacts, products of labor, physical modifications carried out anonymously and collectively, which later generations can't read, for lack of the necessary alphabet.

So Donzelli easily convinced me to make *Venice and the Water* into an independent publication. Is there a more spectacular example of ignoring history anywhere in Italian culture than the story of Venice and her lagoon? If we leave out educated Venetians and scholars who study the city from various standpoints, I believe this page of history is unknown to Italians in general. Few are the intellectuals and men of culture who have an idea of the policy followed by the Republic for nearly five centuries for the purpose of preserving the lagoon. And yet, today this story offers one of the most fascinating *ecological parables* that could be told to the younger generations. It's a chapter in the history of pre-Unification Italy that could have strengthened and nourished the nation's political culture, could have become a popular icon like the episodes of the Horatii and Curiatii, or the Challenge at Barletta, part of the national memory to be handed down with pride. And this case does not involve a minor wartime anecdote, but colossal and eminently practical works of peace, the example of a good government lasting many centuries and preserving a fragile, vulnerable habitat with the consent and the cooperation of all its citizens. How precociously modern, both politically and culturally, is this history of the Venetian Republic: it is without equal in the West! And yet, not a word of it has crept into the national culture, to become a myth or a legend to nourish the imagination and the pride of younger generations. The schools were busy imparting other information, and the intellectuals weren't even aware of its existence.

This model case of ignoring is not incidental, however; in fact, it gives a symptomatic glimpse of the long-term characteristics of Italian culture. It's a chapter of our national experience that deserves to be explored at greater length. In our country, technical knowledge has suffered from being shut away and limited: it has not been able to transform specific knowledge into material for application within a framework of general knowledge and values. Constantly occupied with transforming reality, with making science and human labor into a powerful tool for transforming society, technology has not supplied the language and the myths for presenting itself and becoming part of the country's history. But the root causes for forgetting physical transformations as part of Italian culture don't lie here. Just recall the fate of a great man like Carlo Cattaneo [1801-1869], the intellectual who perhaps more than any other managed to give meaning, value, and historical breadth to the humblest material transformations made in his time. Who in Italy better than he showed what a happy and original synthesis could be made between technical knowledge and humanistic learning? And yet Cattaneo was relegated to the dark pit of the nation's forgetfulness. He never conformed, and he never started a movement. He was briefly remembered recently when the federalist polemic—by now long dead—was revived in a political debate that was short of ideas and in need of some noble garment to hide its nakedness.

In fact, the ignorance under discussion here emphasizes a basic and long-standing part of our national life: the radical separation among the humanistic disciplines in our country. Those disciplines enjoy a definite preponderance in promoting themselves and their value, using communication and writing, having the power to construct identity, general awareness, opinion, hierarchies among the dominant values, and criteria for relevance. Humanistic culture in our country (and the education that has been derived from it) has played a decisive role in shaping generations of men who are absolutely unaware that they are living and acting on the solid basis of a natural and historical reality. That culture is primarily responsible for the experiences and processes of transforming Italy's territory, one of the greatest and least understood social epics in the West, being exiled to a kind of gray limbo, a lesser world, not worthy of attention, to be kept far removed from any curiosity or interest, and above all to be excluded from the nation's heritage of memories.

A natural inclination to give out untested opinions, a predilection for rhetoric and ideology, a constant flight from the humble harshness of reality, such are the signs revealing the stubbornly caste-like relationship of most Italian intellectuals with the society around them. Generally indifferent and alien to the processes that really change the world—the work of labor and technology—they have always preferred self-referential conversation among themselves (on the topic of their own "higher world") or the dialogue with power, above all with the political command centers. Not just to flatter power and use it, of course, but also, in a few cases, to oppose it. And yet, always with the unshakable conviction—a kind of mental archetype, fundamental and not debatable—that the reality principle, the *primum mobile* of every living phenomenon, is the power to command, and its literary expression. Almost as if the only sphere possessing meaning and value were the political sphere, as an ideological narrative, with its rituals and its rhetoric. In possession of words, masters of "discourse," they've chosen that privileged terrain for playing their role, which is much less one of publishing specialized knowledge than maintaining an exclusive, privileged relationship with the universal "languages" and the reality of domination. It's a choice of obvious "economic" convenience. Using the four or five categories of political and humanistic knowledge, it's easy to take part in the daily marketplace of "political discourse" with no arduous scientific effort. It's no coincidence if we still see today, on the scene of our national chatterings, distinguished minds eternally grappling with today's various and everlasting *isms*, wielding old ideologies polished to look new, retelling clumsy stories, reviving things that are dead, creating new alignments and factions and phony battles. They don't retrieve anything worthwhile from the depths of the past, and they're ignorant of what's really happening in the unexplored depths of the present; therefore, they are absolutely barren of proposals for the future.

This history of Venice recounts the experience of technicians, hydrographic experts, fishermen, and woodcutters: an obscure, unknown past, made up of the labor and mighty exertions of individuals, and above all the concerted action of the community. It also recounts a history that is eminently political: the evolution of an exceptional Italian wisdom in governing men and their difficult territory. It recounts and illustrates a patient ability to reconcile differing and conflicting interests and make them all converge, to varying degrees, towards a common purpose. The evolution of this policy displays two eminently modern features: an ability to listen to the knowledge and experience of the technicians, and an ability to translate their advice and their opinions into difficult and sometimes drastic decisions. It's a piece of our past, both humble and unquestionably great, and it contains a kernel of teachings that can enliven our present. Governments that cannot decide, inept democracies, devalue in the eyes of their contemporaries the very meaning of political participation, turning democratic institutions into empty, useless dummies. In an age when decisions are everyday realities in the world of enterprise, banking, and private associations, politics that cannot make decisions sinks to the level of a gratuitous exercise, a pointless ritual for a small separate caste that *thinks itself* and tries to survive as a class. This is one of the most serious threats our country could encounter in the near future. And it threatens to doom the lagoon and the city of Venice to paralysis. Today, around these two realities, a contest is being played out that reveals the capacity of Italian democracy to cope with the environmental dilemma of our time. This case focuses once again on a physical threat of the first magnitude, and the possibilities for response and solution through human action. This chapter of our history, recounting the success with which one government faced up to challenges that are now long past, as well as the limited and controversial achievements of more recent times, may offer a word of hope today in the *possibility* that such a government may again flourish in our country.

Rome P.B.
September 1998

Acknowledgments

I wish to thank Dr. Paolo Selmi, the director of the State Archive of the Republic of Venice, and Dr. Marino Zorzi, the director of the National Library of St. Mark's, who, together with a staff always generously available to scholars, have made my work easier. I also owe thanks to my friends Rita Cervigni and Lucia Moro, of the National Archives in Rome, who made accessible to me, still in proofs, the priceless guide to the Venice State Archives. For recent history, I am greatly indebted to the generous friendship of Franco Miracco and the staff of the New Venice Consortium, especially Monica Ambrosini, who supplied me with documents and illustrative materials.

This essay is one of a series of studies on environmental history funded by the National Council on Research and University Funding (60 percent) and carried out in the Department of Modern and Contemporary Historical Studies, College of Education, "La Sapienza" University of Rome, and continued in the Department of Modern and Contemporary Historical Studies, College of Humanities, University III of Rome.

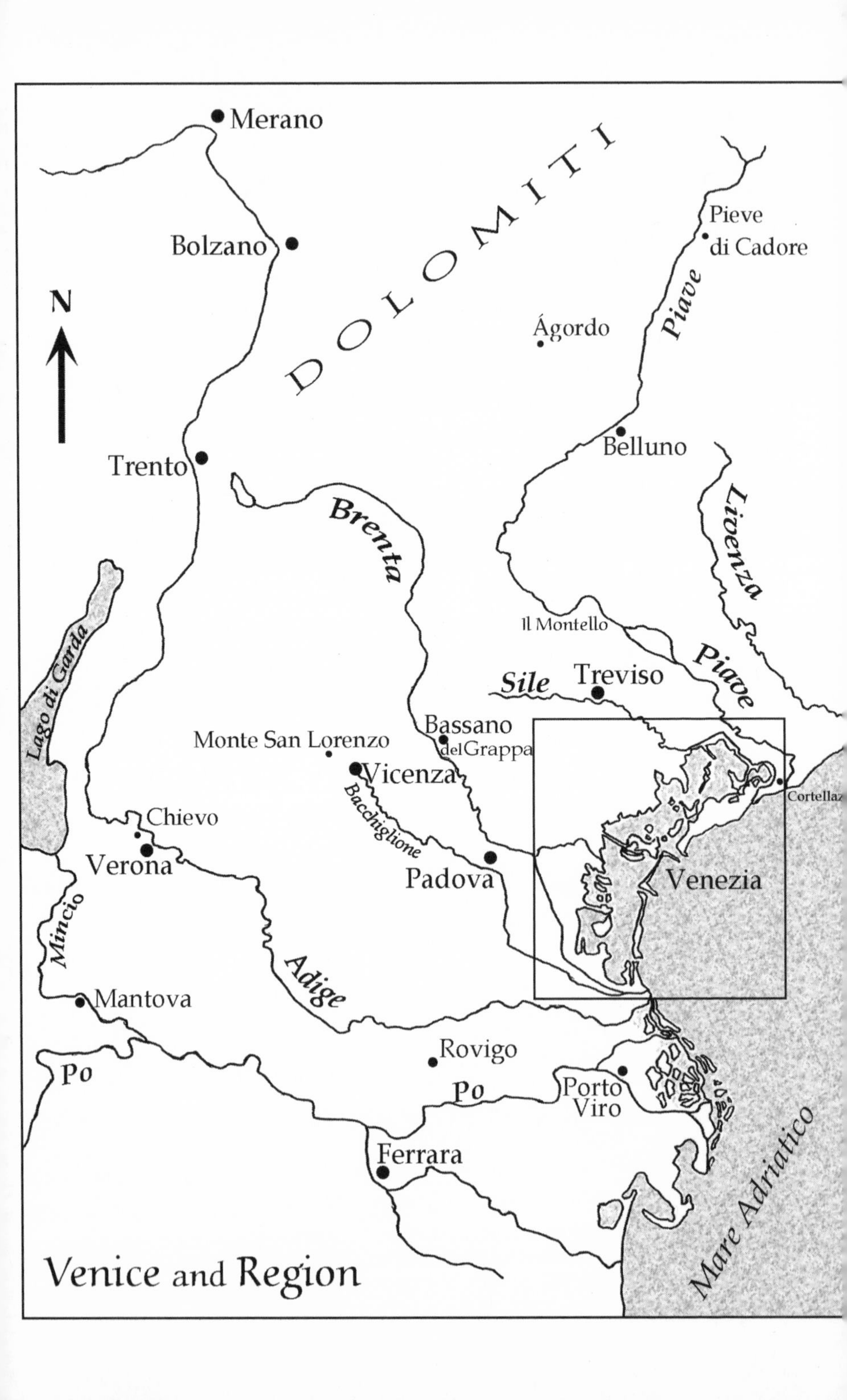
Merano
DOLOMITI
Bolzano
Pieve
di Cadore
N
Ágordo
Piave
Belluno
Trento
Brenta
Livenza
Lago di Garda
Il Montello
Sile
Treviso
Piave
Bassano
del Grappa
Monte San Lorenzo
Vicenza
Bacchiglione
Chievo
Verona
Padova
Venezia
Mincio
Adige
Mantova
Rovigo
Po
Po
Porto
Viro
Ferrara
Mare Adriatico
Venice and Region

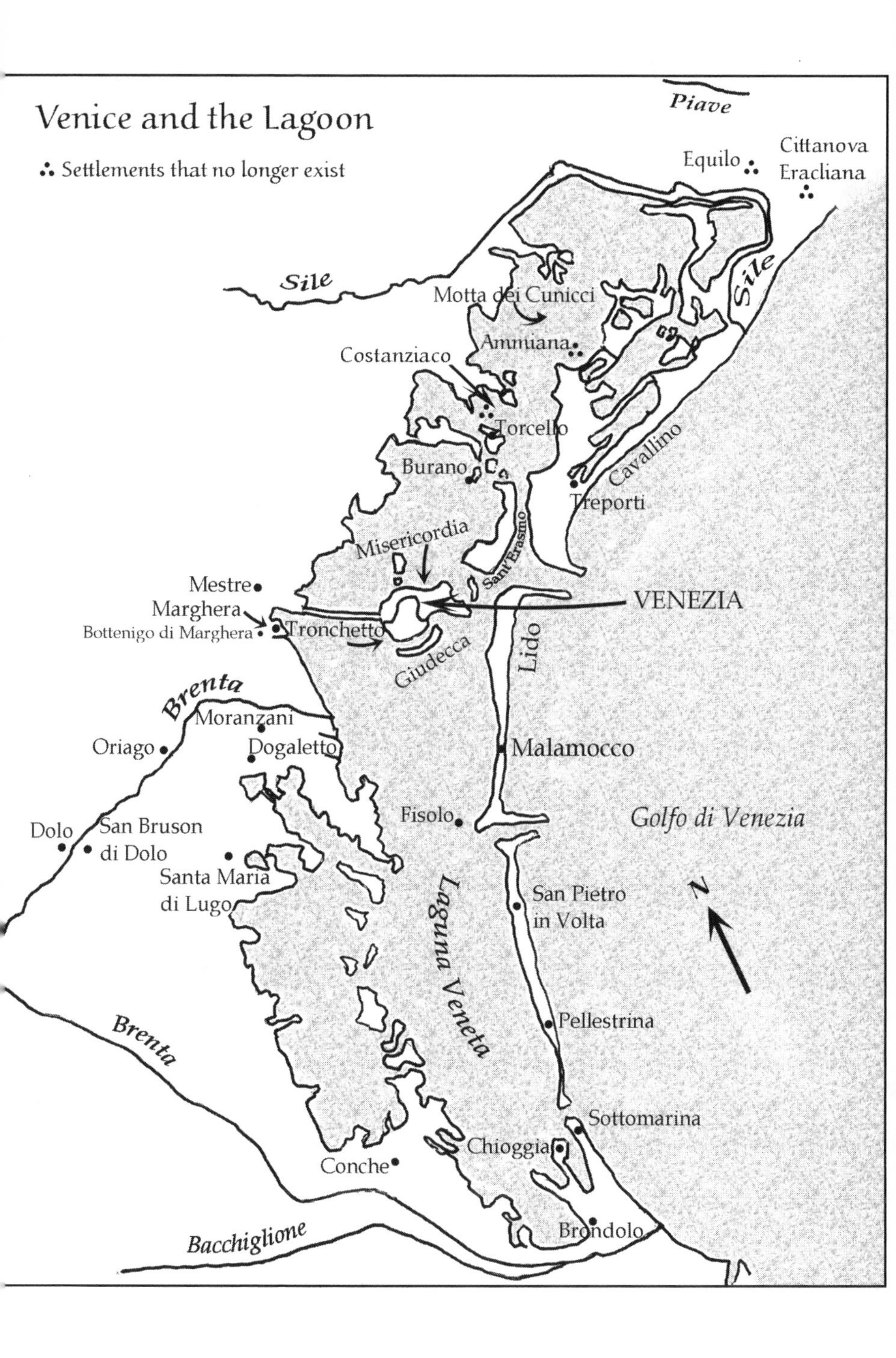
Venice and the Lagoon
∴ Settlements that no longer exist
Piave
Equilo
Cittanova
Eracliana
Sile
Sile
Motta dei Cunicci
Ammiana
Costanziaco
Torcello
Burano
Cavallino
Treporti
Misericordia
Sant'Erasmo
Mestre
Marghera
Bottenigo di Marghera
Tronchetto
VENEZIA
Giudecca
Lido
Brenta
Moranzani
Oriago
Dogaletto
Malamocco
Fisolo
Golfo di Venezia
Dolo
San Bruson
di Dolo
Santa Maria
di Lugo
Laguna Veneta
San Pietro
in Volta
N
Pellestrina
Brenta
Sottomarina
Chioggia
Conche
Brondolo
Bacchiglione

I

The City Under Threat

1. Venice Tells of the Future

Why revisit the history of Venice today, focusing attention once again on her experience, her efforts over the centuries to control the waters that surround her? What's the point of going over familiar facts and phenomena again, since they've been examined and recounted so many times in the vast technical, historical, and pictorial literature? What is there left to add? Hasn't the city's grandiose career as a Mediterranean power been so thoroughly examined and documented that it can be consigned to posterity once and for all? We could answer by referring to a cultural process familiar in the West: every generation rewrites the past, choosing from a huge accumulation of facts the developments and interpretations that meet its needs. So it has been for countless events and periods of the past. Such is the fate of history, changing its perspectives in every age, discovering new sources and rereading the old ones.

But this still doesn't justify revisiting the Republic's past for the *n*th time. Another reason is that Venice's experience is bound up with a cultural reality that is out of the ordinary: she always saw the processes and events in her lagoon habitat in their historical perspective, questioning the past day by day and comparing it to the present, always watching for any sign of change, and deriving hope and advice for the future. That peculiar tradition lasted for centuries and still goes on today, with a program of documentation and reporting that is nothing short of prodigious. Few other cities in the world have been so intimately bound up with the changes wrought by the passage of time, or like Venice have inherited a past wrested from a hostile habitat; few like Venice have faced an uncertain future that must be earned day after day. Venice has always seen herself as special and threatened, forced to detect the slightest change taking place in her amphibious environment. For this reason she has always measured her *passing in time*, her history, ceaselessly accumulating reports and records. In this way she accumulated a boundless heritage of memory that celebrates her glory today, even as it delights scholars. The Venice State Archives are where the outstanding proof is kept, centuries of daily observations bequeathed by the city to posterity.

Today there is a special, deeper reason for recalling this history to the attention of our contemporaries. It comes from the forces of the present, the new historical context posed by industrial societies. Our present situation, our precarious relationship to dwindling resources, our environment that is steadily deteriorating and threatening us, all make us turn to Venice's singular past as to a history that in a certain sense faced our own problems, centuries in advance. Never more than today has Venice spoken to historians and the general public in more up-to-date and universal terms, with her experience in following a course beset with dilemmas arising from her risky and always precarious relationship with nature. Her amphibious position between land and sea, within a lagoon constantly threatened with silting up and other kinds of deterioration, soon drove her to apply safeguards that few other Western cities have ever had occasion to attempt. For this reason, never more than today has Venice given us an example to follow, a model, a strategic context for action that enabled her to win out over the challenges that threatened her very survival. It was a government undertaking, one that has few parallels in European history, one that combined technical knowledge with exceptional financial commitments, and it succeeded in overcoming the threats jeopardizing Venice's future.

This is a history of *success* in managing the environment, a success rooted in rigorous and farsighted government action, in centuries of daily efforts to subordinate private interests to the general good of the lagoon and the city, rooted in readiness to draw a line balancing the economic freedom of the citizens with the constraints imposed by public resources. This history is certainly not easy or idyllic, but agitated by the often unruly behavior of social groups and scarred by internal conflicts and disruptions that were laboriously healed. But no experience is ever linear and triumphant, not even a success. Such descriptions, with no shadows or stumbles, don't match human affairs. Real history is always compounded of contradictions, failures, and recoveries that are sometimes late in coming. As reconstructed by historians, following a fine selective thread through the gray purgatory of facts, there is always, inevitably, some idealizing of the past. This book is no exception. But the philosophy of government followed by the Republic for at least seven centuries of her existence contains valuable elements which it would be hard to overstate, despite their remoteness, the lure of anachronism, or other such facile and unscientific influences.

This is, beyond question, the most original lesson taught us today, at the end of a millennium, by that singular experience. In that sense we may say that Venice speaks to us of our immediate future, more than of the recent past. When the growth of world population, and the depletion of many resources today considered infinite, impose new rules and new constraints on the workings of society, the strategies of the Venetian governing classes can still teach us, or at least inspire us, concerning a problem that will be central for the survival of our democracy: how to preserve individual freedom, the free pursuit of individual interests, faced with the necessity of collectively regulating the essential and limited goods vital for everyone's survival.

2. The Lagoon: A Natural Bastion

Starting in the 14th century at the latest, Venice had to face a fundamental problem in developing and expanding her trade, and in her very survival: protecting the lagoon from constant threats to its natural balance and the process of silting up. The 14th century is taken as the starting point because that's when the sources begin reporting with certainty. But it's long been generally known that managing the inland waters was a responsibility of the community from its founding; from the very moment, we might say, when it became a city. Significantly, Bernardino Zendrini, the great technician and historian of the lagoon, opens his monumental *Memorie storiche* emphasizing precisely the fundamental nature of caring for the waters.

> I could begin the sequence of what I am about to set forth in very remote times, indeed before the year one thousand, since the Republic was always required to control the waters, and with great mastery she made of them the firm foundation of her liberty.[1]

The vast body of water (about 213 square miles) that separates Venice from both the mainland and the sea, accessible through what were then her five inlets, had at least two functions. On the one hand it represented the city's "outer wall," as some contemporaries defined it, defending her against Adriatic storms. At the same time it allowed ships and vessels of all sizes to enter and leave the city for trade, and it afforded them a safe refuge in heavy weather.

The lagoon was the product of an age-old natural process, a collaboration between the same forces that were still threatening it: the mainland rivers and the sea. In the words of Giacomo Filiasi, a scholar and historian of the lagoon, "from Brondolo to Fiesolo, along 32 miles of coast, there was a hornets' nest of river outlets, the outflows of a swarm of rivers, all separate," some coming from the Verona Alps and the Bassano highlands, from Belluno and Treviso, and from the mountains of Friuli.[2] Here were some of the largest rivers in northern Italy, from the Isonzo to the Tagliamento, from the Piave to the Sile, from the Brenta to the Bacchiglione, and from the Adige to the Po, all emptying into the sea only a few miles from each other.

The huge mass of material those rivers carried into the gulf over the centuries—sand, mud, and every kind of detritus—was pushed offshore by a particular current running down the coast and parallel to it, called the "sweeping current" *[moto radente]*.[3] But storm winds and wave action caused part of the material to be deposited on the bottom, creating characteristic sandy barriers some distance from the mainland, called "dunes" and "banks" *[dune, montoni]*. Thus, broad stretches of the sea were gradually cut off some distance from the shore and eventually constituted an enclosed body of water. The land claimed new area and transformed the sea into brackish ponds.

A 17th-century Paduan mathematician considered the origin and preservation of the lagoon to be obviously the product of a "balance," the end result of "two causes, namely the incoming sea tides driving material towards the land, and the rivers as well as the outgoing tides pushing the land towards the sea."[4] As later observed with the breadth of historical knowledge and more extensive empirical data, the phenomenon showed the thousand-year-old imprint it had left on a much longer stretch of the coastline. That mechanism, a great 19th-century engineer wrote, had slowly formed the Venetian lagoons as far as Chioggia, and the lagoons of Comacchio between the ports of Volano and Primaro. There was a time when there were intermediate lagoons as well, behind which stood the ancient city of Adria, called for that reason the City of Seven Seas *[urbs septem marium]*.[5] On the dunes grew the barrier islands of Venice, stretching in a line parallel to the coast, and between them lay the five inlets leading to the open sea: Lido, Brondolo, Malamocco,

Sant'Erasmo, and Chioggia. Over time the seaward passages varied in name and number, as well as in the estimation and the use the Venetians made of them.[6]

Rather than the health of the lagoon, it appears that the first and most constant preoccupation of the Venetians, understandably, was the sea: its periodic assaults on the barrier islands, the inland waters, and on the city itself. The oldest measures taken by the city government, starting in the 13th century at least, involved creating institutions like the Overseers of the Barrier Islands [*Soprastanti di Lidi*], experts responsible for the defense of those islands; in addition, citizens were explicitly forbidden to alter the various natural defenses against the action of the wind and the water.[7] As has often been recorded by several generations of scholars and historians of Venice, this corner of the Italian peninsula was lashed by at least three different winds, Sirocco [southeast], Greco [northeast], and Levante [east]: in alternation according to the season, they sometimes whipped up the Adriatic for days at a time. Those events were natural in the region, but destructive nonetheless. Some of the winds, like the Sirocco that would rage between mid-September and April, often brought "storms that hurl the sea against the barrier islands and the islands of Venice with such force that they seem to threaten a new Deluge." Sirocco storms, Filiasi records, "double the height of the tide and churn the sea to its depths, blocking the mouths of the rivers and dumping terrible rains on the Alps, melting even the oldest and deepest snow in a short time, making the brooks and rivers overflow their banks, and putting all the plains of Veneto and Lombardy under water."[8]

Obviously, storms threatened more than the physical safety of the inhabitants. They also impacted the economic activities of the city in various ways. The salt spray [*spalmeo*] driven by the northeast winds damaged plant life, trees and crops, more or less seriously. Storms could also cover the peasants' vineyards with dunes in a short time, or damage them with the familiar and feared Sirocco. The market gardeners of Chioggia earn a good living with their labors, as a Superintendent [*Rettore*] reported to the Senate on 22 April 1603, provided "the defenses of those islands are maintained, since they are constantly being pounded and wrecked by the unpredictable sea."[9]

As late as the 18th century, when agricultural land had made further gains within the area of the lagoon, it was routine practice to use the mud dredged from the canals to protect vineyards and gardens, as recorded by a contemporary: "The gardeners intending to use said mud by making it into low dikes so as to protect the vines from waves and sprays of salt water in times of Sirocco, and the turbulence caused by the southwest wind, Garbino, to which the vines themselves are always exposed."[10]

Sometimes during storms the sea could rise and flood the enclosures *[fondamenti]* of the salt pans, dissolving in no time the salt that was one of the greatest treasures of the Venetian economy till the 15th century, and one of the most vulnerable.[11] Moreover, unexpected high tides in the middle of the night, often flooding the streets and squares in the fall as they do now, represented a serious threat to the population. They could pollute the fresh water in wells that hadn't been promptly and adequately sealed by the watchmen, who were responsible for coating the covers with clay. But storms themselves, if particularly sudden, produced the same effects. As recalled by a 16th-century water expert, the waters of the sea "in times of storms and Sirocco rise throughout Venice and cause great damage to the whole community, as well as to much property and many wells."[12]

Towards the end of the 16th century, a few contemporaries went so far as to blame polluted drinking water for the spread of plague among the Venetians, an event that in those days was particularly feared. "But who can doubt," asked the treasury lawyer Filippo di Zorzi, "that the plague that afflicted this glorious city of Venice in 1575 and 1576 came from any other source than the rushing flood of sea water that invaded every street and every dwelling, filling the city with excessive dampness, polluting the water in almost every well, and bringing on that evil and pestilent influence?"[13]

As we shall see, the Venetians had prepared specific, if inadequate, defenses against the violence and erosion of the sea: reeds, clumps of tamarisks, jetties *[pennelli]* and breakwaters along the coast. The best protection, though, was afforded by log palisades *[palade]*. They were erected near the barrier beaches, especially at the most vulnerable points, and they were the object of constant maintenance and surveillance. But, as the responsible magistrates insistently repeated, they often fell victim to "the mischief of men," a specific kind of threat to the integrity of Venice, with which we must now become acquainted. "Besides the onslaughts of the sea," records a 16th-century source, "the *palade* are vulnerable to the mischief of men, and since the watchfulness of the Supervisors is never sufficient, on the darkest nights and even during storms a whole installation may be dismantled for the sake of some wood and a few nails, as we have learned to our intense displeasure from the Overseers and the Master of the Barrier Islands."[14]

The sea and its storms represented an almost constant danger to the

Venetians, imposing on them—as recorded by some of the Sages and Executives for Water in the 16th century—the daily "care and preservation of the city's barrier islands, which being always exposed to the assaults of the sea also require constant repair and regular provision," meaning funds for maintenance.[15] But the storms were periodic and recurring events, linked with the seasons and with fate; they were not continuous or progressive. Though calamities were isolated and occasional, they required constant watchfulness and endless expenditure from the public coffers, for the building of seaward defenses. On the other hand, some observers maintained that the force of the sea bringing sand into the lagoon was in fact one of the periodic, and therefore constant, causes of silting in the inland waters, and so a systematic enemy of the lagoon. But this was not a universal conviction. Most specialists were of the opinion that even the violence of the winds, and the tides rising and falling in the lagoon every six hours, constituted "the breathing of this lagoon," as a doctor put it in the 16th century, resulting in a beneficial action of cleansing and renewing the waters, into the remotest coves and canals.[16]

By the 16th century, the prevailing opinion of hydrographic science was summed up in the motto, "A great lagoon makes a great port" *[Gran laguna fa gran porto]*. The sea's incursions increased the area of the inland waters and made them more navigable. Therefore, Venice required openings to the sea, not barriers against it. From time to time the Adriatic became a threatening and destructive force, but it was also a precious source of life for the city's habitat. Furthermore, even those who, partly in error, observed silting in the lagoon—like for example the Paduan mathematician Domenico Guglielmini, at the end of the 17th century—did not draw alarming conclusions for the fate of the inland waters. More "often (at low tide) we see mud flats and marshes near the inlets, than far from them." From this observation "it is argued that the material for silting up is brought by the sea through the channels, but deposited in greater quantity in places where the water starts to spread out, namely on the banks of the channels leading to the inlets." Therefore, it didn't move beyond that area and could cause at most a few sandbars in the entrances to the channels. "But it cannot transform a lagoon into a beach, much less a meadow," our mathematician concluded: there were no precedents of such a nature as to threaten that kind of aftermath.[17]

So the sea had to be fended off in a perennial, unceasing balancing act, but not permanently repelled or shut out. A different matter, as we shall see, from the way in which the magistrates of the Republic considered the long-term action of the rivers. The great streams coming down from the mountains required another kind of treatment. In this case, the enemy of the lagoon appeared to the Venetians as being different in its nature, threat, and potential.

3. The Silent Enemy: Silting

The city faced a very different threat indeed, alarming in a different way, in the evidence of progressive filling in of the lagoon, though it was not as unforeseen, pressing, or radical as the Adriatic storms. Though drastic, silting was not an occasional event but a cumulative process pointing to the future; in the present it showed filling in and destruction as the fate that awaited the lagoon.

The gradual shrinking of Venice's inland "sea," its symptoms being "little water, lots of land, and little movement *[poca acqua, assai Terra e poco moto]*,"[1] was discerned early by specialists and public officials and measured daily in various locations. The process suggested various short-term consequences, mainly for the future.

For a few observers as early as the 16th century, the omens for the future of Venetian waters were bad. According to the hydrographer Fracastoro, observation of the Adriatic coast showed indisputable signs that the land was advancing, and at some indefinite time in the future, that arm of the Mediterranean would become narrow "like a channel." For that reason, he added:

> It seems to me it may be considered necessary and inevitable, at a time that God alone can say, that this lagoon will become a marsh separate from the sea, either because of silting up or the retreat of the sea that fills this gulf, or from both causes; nor do I think that any human power can prevent it.[2]

But that outcome appeared as the far-off consequence of the processes taking place, and however worrisome, it was the final result of a slow geological evolution that was beyond human forecasting. In contrast, a few signs and traces of modifications, highly significant for the future, also had an immediate impact on the quality of the habitat in the lagoon. Changes made in the natural environment and the movement of the water affected the healthfulness of the air. Reeds sprouted here and there, and stagnant water stood in swamps.

The Venetians had been forced periodically to suffer unhealthy influences brought about by changes in the movement of water within the lagoon. That may have happened as a result of a slow silting up that had long gone unnoticed. In 1321 the Great Council, alarmed, reported the effects that were being felt in many parts of the city. Wherever the normal flow and exchange of the waters was impeded, there resulted "great corruption and disease on account of channels *[rivos]*, pools, and canals, silted for the most part, giving off a great stench, especially in the summer season" [Latin].[3]

In a few cases, however, the deterioration could be more or less sudden, taking on the unexpected nature of a catastrophe. Such was the case with the long-term alterations caused in the lagoon by flooding of the Brenta. In 1439, in the vicinity of Oriago, a long stretch of the river dumped fresh water, sand and silt into the salt water of the lagoon.[4] One of the first great hydrographers of La Serenissima, Marco Cornaro, a constant observer of events occurring in those waters, recalled later, in his second *Scrittura [Writing]*, that in the year following the flood, 1440, "there was much fever in Venice."[5] And with the fever the specter of malaria suddenly spread through the city.

The Republic's other brilliant hydrographer, maybe the greatest one Venice had in the 16th century, Cristofaro Sabbadino, from Friuli, saw the appearance of reeds in the lagoon, especially in the downwind part, as a tremendous danger. This development, the great specialist predicted, "will bring so much bad air in the summer that Venice will not be livable; all the cities and places which are unlivable because of bad air, like Aquileia, Cittanova, and Jesolo, were abandoned when reeds sprang up in their vicinity, downwind to the south."[6]

Towards the end of the 16th century there were frequent outbreaks of the feared "plague-bearing mists," as they were called, evil-smelling vapors from stagnant waters in outlying parts of the lagoon. The silting begun by the rivers was often interrupted, creating indeterminate zones that were no longer water and not yet land, which the population anxiously avoided. As Dr. Andrea Marino commented, observing the phenomenon, "not all the filled-in areas *[ammonite]* have become good land, but principally around the lagoon most of them have turned into swamps, dead waters, reeds, and lands that are too wet for farming or dwelling." Aggravated by the fact that the winds, originally a factor in Venice's healthy air, now "bring not only dampness from the sea, as before, but the thick air *[grassezza]* of the swamps and the smells of the dead waters."[7]

Benedetto Castelli, who held the opposite opinion to Sabbadino's a century later when the latter argued for letting the rivers empty into the lagoon, was just as alarmed by the frequent and visible evidence of changes in the lagoon habitat, although they were limited and local. And from them he drew gloomy conclusions for the future. Aside from the silting up of the lagoon, he commented, there was "other damage, and a disorder worthy of the greatest consideration, namely that when the sun warms those mud flats, especially in the heat of summer, it causes them to give off corrupt and harmful vapors and emanations and effluvia which infect the air and could make the city uninhabitable."[8]

In that century such phenomena could be observed mostly in the dead water *[laguna morta]*, the zone nearest the mainland that was covered only at high tide. Here, in 1618, in an attempt to enlarge the area of the lagoon, cuts were made in the banks along the major channels, the so-called Garzoni cuts, from the name of the engineer in charge, who was convinced "that expanding the waters beyond their natural limits and directing them outside their channels was the only remedy for improving the lagoon," as the Sages for Water later recorded in a document dated 15 January 1672.[9] But without the corresponding depth, they observed, the increased surface of the water produced no improvements of any kind, but became the cause of further silting instead; the Garzoni Cuts led "to those sewers of said upper ponds," which remained very shallow and short of water, and on drying up in the summer only created "the worst kind of air."[10]

Fear of a gradual expansion of stagnant water kept several generations of specialists and officials anxious in various degrees. But aside from the threat of expansion, we must remember the recurrent epidemics caused by the worsening cycle of fever. The mixing of "fresh water with salt water," the Paduan mathematician Fra Stefano Angeli pointed out in the mid 16th century, "causes distempers in the air that occasionally afflict this city."[11]

On the other hand, it was not just the natural evolution of the lagoon, brought about by the inflow of fresh water or localized stagnation that caused deterioration and pollution of the water and the air. The city's productive activities themselves and their impact on the waters of the lagoon, not only contributed to the silting up, but in a few cases had immediate effects on the healthfulness of the environment. Such were the effects of "enclosing" fish pens with barriers—woven-reed screens, reeds, etc.—blocking the natural movement of the currents.[12] Or simply throwing into the canals the refuse from workmen's labor: sand, rocks, poles from staging or demolition were secretly dumped into the water.

Such, in the words of some Sages in the mid-16th century, were "the great disorders committed daily in this our city," with the widespread habit "of throwing rubbish and filth into the canals, onto the streets and squares, and into ruined houses."[13]

Paradoxically, in a few cases requirements involving government initiatives were able to make an immediate impact on the lagoon. For example, in 1583 an expert took a boat and went "to see the cove *[sacca]* near the Arsenale, behind the *Monastero delle Vergini*" [since demolished], where the Governors of the Arsenale were requesting permission to fill, in order to expand the site of the establishment. The lagoon, he complained, "should be protected not like life, but like the air," whereas it was being "seriously compromised both by private individuals largely for their convenience"—which was the standard constituting violations—"and for the sake of extending the Arsenale."[14]

Finally, during the modern era farming steadily became more widespread and intensive around the lagoon, with its need for land, levees *[chiusure]* and fences, often such as to create an obstacle to the natural flow of the waters. The reader will become acquainted with this aspect of the issue later. Despite many and frequent prohibitions, the very success of cultivation in fields and gardens dumped rubbish and irrigation water into the lagoon and fostered the formation of swamps. Even towards the end of the Republic, when the greatest problems of living in the lagoon had been solved, the ancient fears and perennial recriminations were still current.

"So the inflows go on," records one of the city's great historians at the end of the 18th century, "and therefore silt *[le torbide]* is rapidly expanding the swamps and the mud flats."[15] Such a fragile habitat was easily damaged by the pressures of economic activities that had scant regard for nature and her limits.

4. Navigating the Lagoon

But as we know, localized pollution of Venice's air was only one part of the more general disruption of balances that threatened the lagoon. Indeed, it cannot be separated from one of the fundamental concerns of Venice's governing classes: preserving the navigability of the inland waters for the city's transportation, and maintaining the depth of the inlets *[foci dei porti]* for seagoing ships and international merchant traffic.

From at least the 13th century on, navigation within the lagoon, in certain cases and at certain points—for example the channel leading from San Marco to the port of Malamocco—could prove difficult because of the silting in or narrowing of smaller channels. Sometimes individuals illicitly reclaimed from the water new parcels of land for farming or building. "Nothing is more harmful to this city of ours," the Sages for Water recorded on 26 June 1531, "than the aggressive filling done for private gain."[1] The transformation of salt flats *[barene]* (ridges covered with vegetation, usually along channels) and mud flats *[velme]* into swamps, and then into land that was then farmed, was more and more visible and alarming. This phenomenon was recorded by officials and experts making daily rounds by boat. Near Sant'Erasmo, during an inspection on 6 March 1573, the Executives for Water observed a "palisade of oak logs" well under way, enclosing an area comprising a vineyard owned by an individual who meant to enlarge the area of his holding. In the same area, the experts assert, "there is a large extent of water and salt flats," and the project underway "to all appearances is meant to enclose a large area, with serious danger of filling in the lagoon." The owner claimed, through "his attorney," ownership of the salt flats. But the officials pointed out the damage that enclosure was causing to the waters of the lagoon overall, and "said palisade was then removed."[2]

Another landowner, during an inspection on 15 March [1531], showed a parcel "dammed off from the lagoon" with levees *[arzeri]* "in which there is an outlet *[scolador]* that drains the water into the lagoon; within the levees on the landward side there are remnants of salt flats." Questioned about the enclosure, the owner argued that without the levees "the Sirocco would drive salt water into the vineyard."[3] Going towards Chioggia in their boats, the Executives came upon a fish pen surrounded on several sides with levees "beside which levees on the lagoon side there is much silting up."[4] During 1580 the Executives ordered the elimination of many levees that were impeding the flow of the smaller channels *[ghebbi]* between mud flats and salt flats.[5] Often the authorities and experts discovered a short time later that palisades were being rebuilt in spite of an order to demolish them. For this purpose they often relied on information from individuals who regularly traveled the lagoon: the fishermen. In a few cases they also turned to other groups of citizens: "This past Saturday a number of peasants were questioned, and they said much filling was being done in those parts."[6]

Efforts to appropriate and transform margins and parcels of the lagoon were constantly observed here and there, and the city's magistrates essentially managed to keep them under control. And yet, notwithstanding all the attempts made by the Republic, starting in the 15th century, to keep large rivers from emptying into the lagoon, the problems of the lagoon and its basin were always on the agenda, despite periodic dredging, inspections, and measures taken to punish abuses by private individuals. The *"Laguna morta,"* stated a contemporary at the end of the 17th century, bitterly feigning satisfaction, "is all being turned into excellent farmland." The writer went on to indicate the various parts of the inland sea where the passage from swamps to farm fields had taken place before his eyes, and he urged his readers to observe all the "evil omens" that could be seen at the time, like "the great swamp below the Bondante, and the diverting of the Dogaletto just a few years ago, that will cause the final collapse of the Port of Malomocchio [*sic*] and the city of Venice herself."[7]

Often, however, the scope of actual phenomena was overstated, and the reporting was intended above all to cause alarm, to exhort the magistrates of the Republic to greater severity. The pessimism shown by the specialists and engineers and observers was very often a *rhetoric of safeguarding the lagoon* that kept the minds of the authorities alert, fostering the general sense—already widespread and substantial—that the lagoon was supreme and not to be interfered with.

Even in the course of the 18th century, when most of the rivers in Veneto had been diverted away from the lagoon and, as we shall see, some of the major problems in protecting it had been solved, silting caused difficulty in getting around every day, and it was a threat for the future. All the more so since specialists and writers very often contrasted the present state of the lagoon with its past condition, when numerous historical records—*e.g.* the passage of galleys in places that were no longer navigable—could prove a greater depth for the inland waters.[8] In this phase, the Sages were constantly petitioned by individuals and local magistrates to allow dredging in various secondary channels or in the vicinity of the inlets, to make them easier for vessels to negotiate. As in the previous century, recurrent silting obstructed the "main channels" *[Canali Reali]*, like the Giudecca Channel in 1688, the "most vital and important avenue for foreign ships entering and leaving," and lesser waterways like Rio di Castello. Here, as a Chief for Water *[Proto alle Acque]* reported on 23 March 1726, the silting was so far advanced "that it keeps even small boats from passing that must travel through the same, especially boats that have to reach the cathedral." The same was true in 1735 for the channels of Rio Morto or Re di Fisolo, so constricted "that public shipping has difficulty turning."[9]

In addition, the normal life of the city, with its daily traffic and teeming activity, also changed the natural rhythm of the lagoon, no different from what's happening today. A contemporary, a treasury lawyer in the Water Department at the beginning of the 18th century, wrote that while the Grand Canal was being dredged opposite the Dogana, "the garbage draining from the city, filling the Rios, the filth stirred up by siroccos dissolving the ground into swamps and salt marshes, and the many ships that anchor crosswise of the canals impede the current and make the waters very slack."[10]

Giulio Rompiasio didn't fail to grasp the basic problem—the limited area of water and modest dimensions of the lagoon—when he wrote, "This body is diseased because it is clogged by silting. The basic [cause] is natural, lying in the behavior of the lagoon because of its makeup, comprising many different and opposing water currents, large numbers of cross channels with various divides *[Partiacqua]*, apart from the main roads to the inlets. In addition, the lagoon is of limited dimensions in length and breadth, so that the waters cannot spread out during the 6 hours of rising tide in the sea and as many of ebb."[11] A second cause was still, according to a centuries-old tradition of reports by specialists and magistrates appointed to survey the waters, men's "malice in their accursed infractions of the laws governing the lagoon." We shall return to this point.

The persistent pessimism of the authorities and observers about the fate of the lagoon at that time can also be explained by the condition of the inlets. If Malamocco had certainly been improved by the 17th century, freed of the silt that threatened to obliterate it entirely, such was not the case for the other inlets from the Adriatic.[12]

For a long time, and until the early 19th century, the inlet of San Nicolò, or Lido, had been considered impassable for any but small boats: it was a great loss, because that gateway to the lagoon had long been the port of Venice par excellence.[13] But the silting up of the inlets or their deteriorated condition, more symbolically perhaps than in other parts of the lagoon, represented the threat most dreaded by the Venetians: the loss of navigability from the Adriatic and hence the end of trade, of wealth, of power. Ever since the late Middle Ages, when the city emerged as a trading center for the Mediterranean world, every sign of change in the physical conditions for navigation—the waters of the lagoon—had been observed and experienced as an immediate or a long-term threat to that power, to that unique achievement of centuries, never to be repeated. The lagoon was the support, as contemporaries were to repeat at various times, of "Security, Liberty, and Health."[14]

The future of Venice was burdened not only by the present physical circumstances and their possibly dangerous evolution, but also by the memory of the past, even the most ancient past. Along this end of the Adriatic, to the west as well as the east, nature had always undermined the urban outposts along the coasts. "Silt invades coastal cities," recalled Predag Matvejevic, referring to the eastern shore of this enclosed arm of the Mediterranean: "Not everywhere do they have a favorable destiny."[15] But in the common memory many histories are marked by a sinister fate. Islands like Monte San Lorenzo, Motta dei Cunicci, Santa Cristina, Sant' Adrian were abandoned because they were invaded by reeds and poisoned by malaria or made inaccessible by relentless silting. The tales handed down and the traces still visible of abandoned sites, urban centers, places once inhabited by people, told not just of swamp formation and silting up, but also of sinking and disappearance beneath the waters. The memory and records of the northern islands of Equilo, Costanziaco, Ammiana, the lagoon of Cittanova Eracliana, and even the partial submersion of the port of Malamocco in the 12th century, handed down a similar tale of decline and abandonment. As it appeared clearly from the 18th century on—according to some, at least from Sabbadino's time—if the lagoon tended to silt up because of natural processes of deposit linked to the transport of materials from rivers and even maritime sources, now it was being threatened by a much more serious and opposite phenomenon: the rise in the level of the sea; or perhaps more accurately, the slow but relentless sinking of the whole territory of Venice. This is the same problem, as we shall see, that faces the city in our time and makes its future uncertain. At the beginning of the 19th century, when the threat from the rivers had essentially been turned aside and basic problems of defense against the sea had been solved, it was precisely this condition that moved Filiasi to say, "if this goes on, in a few centuries it will surely be necessary to rebuild Venice on top of herself."[17]

On the other hand, not only were the threats to the health of the lagoon numerous and caused by different factors, but they could not all be judged to be equally serious. Interpreting them was entrusted to the opinion and judgment of experts; their decisions were subject to error and almost always contested or harshly debated by other experts. Undisputed scientific certainty was not something that could reconcile the opinions of experts. Indeed, such a condition still obtains even today, regarding events and processes that threaten the whole planet. In one case, for example, even one of the most authoritative 16th-century hydrographers, an energetic defender of the inland waters, wound up advising that certain flats be filled, to speed the currents near the cove *[sacca]* located "behind la Misericordia." And the Sages for Water must have been astonished when they read Cristofaro Sabbadino's assertion: "I do earnestly advise filling in that cove, for as I have said, it will not at all be harmful to the lagoon."[18]

On the other hand, the lagoon often displayed variations for which the most attentive observers could not account. For example, silting was not always present; sometimes the opposite took place. According to the testimony of some experts, for mysterious reasons even salt flats could wind up under water. An expert observed in 1551 that where there had been "two large salt flats, now there is nothing but water; I have seen with my own eyes what I have stated, and in many places I have cooked fish."[19]

Of course we can't check the truth of such a statement, which concerned other salt flats as well. But its truth is secondary to the indirect testimony it gives us: the essentially mysterious behavior of nature in the lagoon. Filiasi himself, writing after generations of historians and specialists, emphasized the mysterious behavior of the inland waters, displaying "a great variety of more or less frequent events," and went on, "Both before the rivers were banned [from the lagoon] and since then, there's been so much back-and-forth that the history of our waters has been shrouded in fog: taken in isolation, the project is cited by every side in support of its own opinions and theories."[20] Nature still remained hidden, changeable and ungraspable, so interpreting her signs separated men into parties and factions as if by an evil spell.

5. Present Dilemmas and Future Mystery

Such, then, was the singular and modern situation the city found herself in: her horizons were uncertain and fraught with danger, but the danger did not come from an unavoidable fate or from sudden, unforeseeable disasters. The threats were developing in step with the long-term evolution of conditions, towards a future destined to cumulate and enlarge events that were already visible but still circumscribed. These events were not only supposed to be interpreted by the technical experts, with their controversial prophecies, but they quite often resulted from human intervention, the very solutions put forth by the hydraulic engineers and the authorities, to remedy imbalances and disorders long observed.

As Contarini well knew, writing in the 17th century, "It is easy to commit huge errors that do harm to the community, and the public treasury cannot suffice to repair those great disorders, which may last for centuries."[1] Moreover, the outcome of some of the solutions adopted was not immediately visible, as was often the case with diverting rivers from the lagoon, a technical decision that divided generations of experts and writers. There was a long delay before evaluation was possible; this fact alone increased uncertainty about the decisions made, and fueled controversy. As Zendrini wisely remarked, this was "in no way surprising, for men always condemn actions which do not immediately produce results appropriate to the goal for which they were undertaken."[2]

Moreover, interventions involving the water didn't take place inside the retorts of a laboratory; they resulted from government decisions made on the basis of expert advice and long evaluation, and sometimes they entailed large investments from the treasury, money that was taken from other government duties. Therefore, every choice was almost always hard and burdensome in several respects, and great was the risk of not solving the problems at issue, of unintentionally upsetting the fragile balance that already existed. The status quo is not something to be changed lightly. As a Sage for Water complained in 1672, "it is very dangerous to think that the greatest evil is doing nothing, while the worst of the worst is taking action in error," at the risk of "adding disorder to disorder and harm to harm, causing everything to collapse all at once, and very soon."[3]

Likewise, our Bernardo Zendrini again recommended "going very cautiously about changing things in any way that may tend to upset the inner balance of the currents in the water, especially the ones that nature herself made into a system, for she will always know more than men, however much theoretical and practical knowledge they may boast."[4] And yet the possibility of error, however great the risk, could not be a sufficient reason for not intervening or attempting a possible remedy. Standing by and watching the spontaneous processes—or ones resulting from earlier interventions—while they degraded the water was certainly not acceptable to the Venetian authorities, which had been established for the purpose of acting to correct the observable phenomena and trends.

Nevertheless, for several centuries a general aura of uncertainty hung over the entire destiny of the lagoon and the city: how much time would it take for the waters to be completely filled in? How many decades or even centuries would pass before Venice truly entered her certain and irreparable death agony? Here again, of course, there was debate or at least divergence of opinion. Even those who were convinced that Venice would inevitably be silted in—though it might not mean the end of the city—looked realistically on the successes obtained, especially the diversion of the Brenta; but they did speak of the possible threat, the sudden collapse lurking in the course of a slow evolution. In his statement, a 16th-century contemporary, Dr. Andrea Marini, shows remarkable power of observation: "I don't mean that the present lagoon cannot last a long time, although the dangers surrounding it are very grave, because since the Brenta was turned aside there have been slight losses in some places and great gains in others: many important channels were shallow and have grown deeper. Nevertheless the lagoon's enemies are great and powerful; though they are slow to take offense, the offense will prove fatal in time."[5]

Of course the gods have given to no one the ability to prophesy how long the future of Venice's waters will last, and this has made for uncertain relations with the younger generations as well. A problem that's so typical of our time was already fascinating the great minds seeking to discover the destiny of the Republic. Some, like Benedetto Castelli, favored letting rivers empty into the lagoon; he stated with striking clarity the dilemma of relations with the coming generations in the near and distant future, and the choices that he considered necessary in light of their future benefits.

> And I have demonstrated that if the Sile and the other rivers had been diverted as intended, the lagoon would have practically dried up in a few days, and the sea inlets would have been lost, along with other disastrous consequences; but on the other hand, even if we admit the silting, we can say with all probability that it will only take place over the course of hundreds and hundreds of years. It does not seem to me wise counsel to adopt a resolution now and embark on a course of action to secure a benefit that is most uncertain for the sake of those who will come many centuries after us, while doing certain harm to ourselves, and to our children living and present.[6]

In this case Benedetto Castelli was mistaken, technically, about the diversion of the rivers. The centuries have shown that choice to have prevented the silting of the lagoon, maybe even contributing to the opposite phenomenon, the later rise in the water level. Besides, he didn't realize what different time scales he assigned to the processes of deterioration—"a few days" for the absence of rivers from the lagoon, and "hundreds of years" for their continuing flow—although his point of view was authoritative, it was still just his own. And yet he went to the heart of a major problem: what our relationship with the future should be.

It was uncertainty as to the duration of events, the time necessary for their development, the scale of expectations, and the degree to which calculations had predictive value, that led to essentially short-sighted positions like his, as well as other men's extravagant and utopian positions. And just that state of affairs—in the presence of real and growing damage observed by generations of specialists and authorities—finally created in "public opinion" among the Venetian upper classes, specialists and observers, a sort of *rhetoric of the end* that lasted for centuries. There is no text, manuscript, essay or memoir concerning issues of water that doesn't mention the risk of irreversible damage to the lagoon and hence the threat of destruction hanging over the city. Similarly to what happens in our time—events are reported over and over, but their evolution seems difficult to forecast—if their consequences are delayed or do not occur, the events tend to take on a patina of unreality. A clear disproportion eventually comes about between the dreaded risks as they are repeatedly, often spectacularly pointed out, and the actual outcomes, even their very timing. Thus it is that criticisms and denunciations take on the appearance of a kind of ritual, the mechanical repetition of prophecies that are ignored when too much time goes by. The rhetoric of alarm is invariably replaced with the *rhetoric of disbelief.*

But in partial contrast to our present situation, Venice was well aware of the threats facing her, not only because they could all be observed and verified locally, but they could be documented in the past as well. As we have briefly observed, it was in memories and landmarks, from written documents and ruins—buried or submerged—that the inhabitants of Venice found the most threatening models of their possible future. The future, as a possible scenario for the coming years, contained the events and the processes of the past. For the Venetians, therefore, history was very substantial material for predictions.

It has been observed concerning other European peoples faced with a daily threat from the water, the Dutch for example, that for them the terrible experience of periodic floods took on a deeply religious character. Destructive events forced the populations of those lands to rebuild over and over again, with tremendous labor and expenditure of resources, a habitat devastated by the sea. That religiosity, so necessary to sustain the spirit of peoples often overwhelmed, naturally and powerfully gave rise to the Protestant Reformation. "No one who had come through the floods," the historian Simon Schama points out, "could fail to grasp the distinctive meaning of the *beproeving*, the trial. The constant testing of faith in adversity was thus a formative element of the national culture."[7] The survivors were the chosen, who had the obligation to rebuild their cities. And unlike the Venetians, Schama continues, "whose historic mythology gave them a genealogy of immemorial antiquity and continuity," the Dutch "had irrevocably cut themselves off from their past and were therefore obliged to reinvent it, to heal the wound and rebuild a sound body politic."[8]

In this divergence between two populations faced with similar problems since their origins, it's not just the different course of their history that plays a significant role, nor their different religious and cultural experience. There was also a difference in the kinds of hardships and threats they faced, which led to a different spiritual attitude, and a different relationship to the past. The periodic floods in the Low Countries were unforeseen and unforeseeable—similar in this respect only to the storms that struck Venice from time to time—the causes being of course inexplicable, like any act of God. The past which the Dutch tended to recall, to give cohesion and strength to the national awareness, was a moral one: the examples of self-sacrifice, the ability to fight, and the will to rebuild that past generations had shown on the occasion of past floods. For the Venetians the past was not quite the same thing. They certainly had a mythological and exemplary relationship with their past. Displaying and celebrating their greatness in centuries past, the nobility of their origins, the splendor of their achievements, their internal political stability, and the superior wisdom of their government, all these were frequent intellectual exercises over the centuries. They served to reinforce collective identities, and they obviously gave the leading classes and the established political power an extra measure of consensus, almost a sacred halo.

But the Venetians also had a strictly nonreligious connection with the past, one that we may term purely technical and modern. As already emphasized, the threats to the integrity of the lagoon did not come from periodic and unpredictable events but from progressive, observable phenomena, to be interpreted in the light of scientific or at least empirical knowledge. Choices had been made in the past—diverting a river, opening a channel, enclosing a fish pen—and the present could begin to evaluate the effects. The previous decades and centuries, therefore, gave not just the proof—a submerged island, a filled-in swamp—of what Venice could become in a more or less near future. They also testified to human error or successful choice, displaying before the eyes of contemporaries the consequences of actions taken by their predecessors.

The present and future of Venice and her waters, so closely bound up with the past, were therefore much more obviously the result of human activity, of choices made, of solutions adopted, of means employed. To countless generations of specialists and government officials concerned over the fate of the lagoon, the present and the future of Venice appeared much more subject to the behavior of the population, the commitment of the authorities and the investments of the state, than to fate or the irresistible force of nature. The past was not a myth, but history: it was human action, to be observed and verified.

We have seen, and we shall see again, that one of the main and constant threats to the inland waters of La Serenissima came not from the sky or the sea or the rivers, but from men, from the individual interests of productive workers: peasants, fishermen, boatmen, homeowners, the private citizens whose economic activities in pursuit of their individual interests often conflicted with the natural and collective requirements of preserving the lagoon. This is another reason why the Venetians could only have a strictly secular relationship with their history and an absolutely open-minded, empirical view of the present and the future. As in our own time, both were seen by everyone as intimately dependent on the interests, the behavior, and the material choices of men.

II

Scarce Resources, Renewable Commodities

1. Fresh Water and Brackish Water

The lagoon, as we know, was not just the "fluid plain" of Braudel's old maritime image, used for transport on ships, boats and rafts. It was not just the place of transit for vessels bearing the goods that an eminently merchant city put in circulation every day. Before it was the scene of complex trading economies, it was first of all home to a growing number of citizens needing healthful air and drinkable water. By the early 14th century Venice already counted more than 100,000 inhabitants scattered throughout her islands, and was thus one of the largest cities in Europe.[1] In 1563 the city had over 165,000 inhabitants—the total population of the Mainland Dominion was over 1.5 million—and was second in Italy only to Naples. Throughout the modern era, the Republic of St. Mark remained the most populous state in all of northern Italy.[2]

It is easy, therefore, to understand the concerns expressed in several circles about the pollution of the air by such a large human settlement, one accustomed to enjoying the healthful conditions offered by a seashore location and constant breezes. Even drinking water, derived from rainwater since the beginning, was a scarce resource in that setting and soon became inadequate to the demand. The "Venetian well" was plainly a great technical achievement, an original type of dug well still to be seen today in squares and courtyards throughout the city. Especially Along the sandy shores of Grado, Jesolo, Caorle, and Lido, wells were dug into the dunes and on down through the layer of clay on which Venice is built, finally reaching fresh water. Apparently this scheme met the city's needs for a long time, because not until 1611 did it become necessary to supplement these wells in order to meet new requirements. A canal was then dug, the Seriola, taking water from the Brenta: the water was collected in suitable containers and loaded onto canal boats *[burchi]* near Moranzano, then brought to the city, to be emptied into wells both public and private.[3]

As can readily be imagined, the city authorities took the greatest care regarding such a scarce and precious resource. The construction and restoration of wells—which Venice continued to use until the water main was built in 1884—was the purview of magistrates as powerful as the Overseers of Salt and the Deputies for Road Repair. In the late 14th and early 15th centuries, government control became stricter and more systematic with respect to hygiene, with supervision assigned to the Overseers of Health; engineering and construction were the task of the Overseers of Public Works *[Provveditori di Comun]*. Routine use and maintenance of wells was entrusted to "district chiefs" *[capi contrada]*, who kept the keys to the covers and unlocked them each day "when the bell rings." Around them gravitated a small population of porters, parish employees *[piovani]*, and water peddlers who cleaned and guarded the wells, and plugged the drain holes with clay when the tide threatened to rise higher than the well curbs.[4]

If fresh water represented such a scarce and fragile resource—as well as a commodity—and a vital requisite for human life in the lagoon, salt water was no less precious, as the source of salt for Venice's meals and for her economy. By the Middle Ages, as recorded by the leading historian of this "industry," Jean-Claude Hocquet, salt had become "a source of incomparable wealth."[5] Salt soon became the basic product in a self-sustaining economy that was also, of necessity, dedicated to trade. As a city with no land, at first Venice had to search the lagoon for sources of subsistence as well as goods for trading outside. After hunting and fishing, salt soon became one of the most important items provided by the waters of the inland sea, in this case not just a valuable food item for cooking and for preserving fish, but also a surplus commodity for export to other settlements, in exchange for other commodities and cash money.

The salt industry in the lagoon generally involved huge investments, targeted most of all at creating protective works and the artificial setting for the salt ponds. Dikes had to be built as defense against the sea, channels had to be dug, and a large area had to be leveled as a site for the ponds *[fondamenti]*. These were large-scale works that could reach a perimeter of 4,100 feet in the case of the Cona da Corio pond, or the Lagoon pond with 5,250 feet; and in essence the dikes defined the scale of the operations. The pond was divided in turn into a certain number of pans *[saline]*, from 12 or fewer to 50 and more. The salt pans, which represented the operation of a family of salt makers, tended in turn to grow larger. They were long, narrow rectangles divided in turn into smaller compartments called evaporators *[cavedini]*. A salt pan could number between 20 and 40 evaporators, the larger pans being as long as 460 feet.[6] It appears that in the 13th century, the golden age of Venetian salt, there were as many as 119 salt ponds scattered all around, and their characteristic whiteness was a dominant feature of the lagoon landscape.

Salt production was significant financially, bringing large revenues to the treasury for centuries and supporting one of Venice's richest and most powerful authorities, as we have seen: the Overseers of Salt. In 1428 this body took over from the salt producers of Chioggia and essentially controlled the salt trade between Venice and the outside world.[7] At the same time, as already mentioned, salt making was one of the most fragile economies, exposed as it was to the hazards typical of the lagoon. Sudden storms or flooding by sea water injured it periodically, while more drastic changes in the lagoon, whether deepening or silting, wiped it out completely in some locations.[8]

Until Venice found it preferable to import salt from Cyprus or Sardinia, Sicily or the Balearic Islands, salt making remained a typical economy for that setting, conforming perfectly to the logic and the processes for sustainable use of existing resources. Protecting sea water from all kinds of pollution or alteration and enclosing it periodically for controlled evaporation and conversion into salt was one of the most rational and systematic uses of the renewable resource par excellence represented by the sea. If we remember that for several centuries the Republic built her international trading fortune on this resource, we can better understand the political wisdom of her authorities in defending the lagoon as a habitat.

One scholar has rightly defined the Venetian lagoon as a series of suburbs grouping the economic activities that under the old regime usually crowded the outskirts of cities. Indeed, the Venetians' readiness to make use of every resource found in the lagoon, adding value to commodities available for transformation, tended to make Venice into a huge industrial zone, where land was "built" for cultivation, reeds were used for fish weirs, clay was gathered for making bricks, sand was dug for making glass, and so on.[9]

Among the "industrial" activities pursued in the waters of the lagoon—although like the salt ponds they did not survive the transformations of the Venetian economy—water-powered mills played an important role. Present by 982, the first year for which records exist, they represent one of the most typical expressions of Venice's economic outlook, a tendency to see the water as a lasting and varied source of wealth.[10] The mills, variously called *"molendini," "aquimoli," "sedilia,"* etc., were located on the various currents within the lagoon: the tides, and the continuation of river currents beyond their mouths. There were two kinds of mill: stationary and movable. The former were located on islets or on the banks of channels, and the latter on broad flatboats called *"sandoni,"* which could be moved wherever the current was strongest, like floating mills on rivers elsewhere. They made use of a hidden resource of the inland sea, however limited and subject to unexpected variations: motive power, a relatively rare factor in the economies of the old regime. The mills were just one of the many economies representing the productive use of the lagoon through light equipment that was adaptable to the existing environmental balance.

We cannot say why mills disappeared from the landscape of the lagoon. The presence of a mill was last recorded by Marco Cornaro in 1440. Nor do we know if it was a consequence of diverting the rivers, which eliminated the strong freshwater currents from the lagoon, especially near the river mouths, or if mills were outlawed in order to protect the free flow of the waters. According to Filiasi, the mills must have become very numerous as time went by, since at a certain point the millers enjoyed the same degree of consideration as wine growers or market gardeners, almost as great as the fishermen of the lagoon. Such widespread use of flour mills, if only because of the technical requirements of locating them, may have contributed to the gradual degradation of the existing environmental balance: "Enclosing vast tracts of the lagoon with dikes creating many partial basins, sometimes forcing the waters out of their course and into the basins, and even filling tracts of swampland and building shelter for the millers and storage for grain and other things, as well as doing other things unknown to us, they could only bring about disorders in the lagoon itself." But Sagredo remarks that no express prohibition by the authorities has come down to us, so probably after Venice's conquest of the mainland in the 15th century the millers naturally moved their stones to the banks of rivers in Veneto.[11]

A further reason for doing so was the possibility of war, an ever-present menace then, even though Venice succeeded for centuries in keeping it far from her waters. As a Sage noted on 3 March 1575 concerning the mills along the Bottenigo, in case "any danger of war may befall this Republic," the mills on the river "will never lack water, any more than do the ones in Dolo."[12]

2. Hunting and Fishing

Since the beginnings of the city, or rather since the first settlements in the lagoon, in order to survive the population made fullest use of its most abundant natural assets, fish and birds. In the Ducal age, fishing and hunting were still the trades that supplied most of Venice's food. The two activities were long combined and could be pursued in the same places, since hunting almost exclusively involved shore birds that migrated through the lagoon or were present year round. In the swamps and marshes among the hummocks, salt flats, and channels, "going hunting for meat as well as fish" [Latin] was customary and widespread.[1] Sometimes it represented a kind of seasonal division in the use of the waters and swamps: fishing was for summer; while in winter, when migrating birds stopped in the bushes and swamps, hunting intensified. Ducks and widgeons were the prevalent quarry of Venetian hunters, who used the *bote*, an open barrel set into the mud, often with live ducks for decoys. Hunting was not limited to birds. Until a certain time, it was possible to take boars, stags, wild goats, and hares, a modest range of game which was caught using a variety of devices as time went by: nets and snares, birdlime and muskets.[2]

This type of productive activity, intended to meet the food requirements of the population, was also controlled and regulated by the Venetian authorities, who also encouraged the rational use of the resource, at least as regards fishing. In those times it was of course difficult to control an activity like hunting, which provided an important change from the prevailing Venetian diet of fish. At various times the authorities tried to intervene, especially in the distribution phase, to prevent the spread of indiscriminate hunting and its destructive consequences. In the mid-17th century a proclamation of the Old Court *[Giustizia Vecchia]*, which then governed fishermen—who also hunted in the lagoon—described illegal sales practices widespread among "peddlers from Murano and Burano, who usually sell in various places, especially the portico of the New Buildings *[fabriche nove]*, where they set up their own shops selling poultry illegally, harming and gravely compromising the prosperity of very many people." The proclamation was aimed at unauthorized sale in improper locations, and it ordered: "Let no Peddler or any other person not listed as a Master in the Guilds of this city authorized to sell, dare in any way or on any pretext to retail large or small quantities of Coots or any other Birds, cut up Geese, or any other victuals subject to the courts of Their Most Illustrious Lordships, under the portico of the New Buildings, or in the streets, or in any other place in the City."[3] The city's courts tried to regulate hunting itself, in an effort to prevent a year-round activity that kept the game from reproducing. In 1751 hunting was prohibited for the period starting with Lent and extending through July.[4]

Fishing was of course the economic activity that best lent itself to rational and, to a degree, programmable management of individual enterprise. As early as 1261 fishing was subject to the control of the Old Court, which mainly oversaw matters of food supply, cleanliness of selling places, and as we shall see, the standards, schedules, and equipment for fishing.[5]

Fishermen were organized in communities called "Brotherhoods," with a chief elected periodically; he could require fees from the boats of his district, and he enjoyed other small privileges. The Brotherhoods had codes of laws and customs called registers governing the life of the guild and regulating practices in catching, delivering, and distributing the fish.[6] Fish selling was the province of venders in the guild of Fish Merchants, whose practices were set down in the Fishermen's Register issued by the Old Court in October 1227. They were required to sell their fish in the place established by law, "at the Pole" *[al Palo]* in the traditional phrase, meaning near a mast standing in the marketplace, either at St. Mark's or the Rialto. This was intended to prevent monopolistic buying, price speculation, and fraud or adulteration of the product. Therefore they were forbidden to buy merchandise not sold by recognized fishermen, to form associations and companies of more than two partners, to open stalls other than their own, and to raise the price of fish. These were practices the Venetian authorities constantly tried to repress, branding them "harmful to all" on the part of men who "had more regard for their own greed than the common good."[7] Towards the end of the 17th century, the Old Court had to issue yet another proclamation forbidding citizens to go out to meet "any vessels *[Tartane, Brazzere, Barche, Battelli]*, or other craft bringing fish to the Metropolis, or once a vessel has arrived there, to rush on board to buy or take fish by force from the masters of said boats, before the fish has all been legally unloaded into the public fish markets."[8]

Control of fish buyers and sellers, which was always very strict for fiscal reasons as well, also involved fish brought from outside the lagoon, which became more and more prevalent in modern times as the local product became inadequate. A regulation of 1680, for example, stated expressly that merchants buying eels from Comacchio must fill out printed permit forms and register "at the Pole" the merchandise they bought.[9] Again regarding sales, the Venetian authorities regularly kept watch on prices and intervened to control them. In the mid-18th century prices were set "every two months in consultation with the fishermen, the Growers, and all others dealing in fish." This was part of the government's overall purpose of "keeping the venders of all foodstuffs moderate in their prices" by means of "regulations governing retail sales, with the salutary provision that prices shall not be altered by anyone on any conceivable pretext, but shall be respected scrupulously for the relief of the poor and the aid of all families."[10]

Here was an early example of "social pricing" of fish, in an effort to strike a balance between the demands of producers (the fishermen), the profits of dealers and sellers, and finally the broader, more general interest of the consumers. Between those offering a basic commodity and the consumers a mutually satisfactory agreement could be reached. Here was another field in which, by regulation, the Republic tried to turn the spirit of individual free enterprise towards goals of a general nature, blending the pursuit of private gain with the meeting of public needs.

3. "Planting" Fish

This surveillance, and the constant effort made by the Venetian authorities to subject private economic activities to the requirements of public service in such a critical area as the city's food supply might at first glance be mistaken for a typical old-regime policy aimed at controlling food stores for obvious reasons of tax revenue and public order. Perhaps the concerns of the city's ruling classes were not unlike those of other authorities—regional and national, Italian and European—that blocked free trading in grain in order to minimize the frequent shortages that tormented pre-industrial societies. So those regulatory initiatives partly recall a familiar situation under the old regime. Today, however, regarding that phase of Western history, we tend more and more towards a position that is less rigidly modeled on the interpretations and values of the capitalist and industrial society that replaced it. The destructive and self-destructive capacities of the latter lead us today to quite a different judgment of the mechanisms for social regulation in pre-industrial societies that were once denounced as irrational obstacles to the brilliant prospects of development. In that respect the Republic offers an extraordinary lesson: here it was once possible to combine competition in international markets with protection of the society and the territory. In an unusual set of circumstances, Venice found herself playing an absolutely "modern" role among old-regime societies. For centuries, unlike other states, she did not have the problem of expanding her home territory, opening new land: in short, producing development. Her economic goal, as concerned the heart of the city—the lagoon—was conservation: keeping men and the processes of natural evolution from altering the existing state of things. Her whole effort, concerning the physical place where her power was settled, was to keep it as it was. While she was becoming an international economic power and behaving like one, the Republic was working out and practicing a policy of *conservation* to maintain and restore the natural balances that made living in the lagoon possible.

For that reason too, the opinions of the city's magistrates have a more ambitious profile than can usually be observed in old-regime societies: the whole philosophy of the Venetian ruling classes seems to have been inspired by broader, already modern views. As we shall see, the problem of fair, controlled selling of fish was in no way considered in isolation from "production." Distribution and sales were not just a matter of food supply, unrelated to the way the fish were caught and the seasons, places and times of reproduction.

The fish reaching Venice were caught within the boundaries of the lagoon, some in the open sea, and later some were imported from Comacchio, Ferrara, and Istria. But Venice's fishermen were at their best working the inland waters, and their catches provided the population with food for centuries. In the lagoon it was possible to practice what was called "nomadic fishing," not bound to specific places, using various small boats *[tartanella fissa, tartanella ciara;* or especially from Burano the *traturo da strazzin;* from Chioggia the *bragoto* or *bragozzo]*.[1] Nomadic fishing used nets systematically, some of them large *[paranzelle lagunari, bragona, bilancelle, cogoli,* etc.], some of them still used today, and all duly authorized by the magistrates who supervised the fishery. The mesh of fishing nets was regulated, and it was constantly checked for size and weight against standard nets kept by the Overseers of the Old Court. Every net cast into the water had been inspected and sealed, in an effort to dissuade the many counterfeiters "who with various fiendish inventions consume every kind of fish, catching them with endless devices that have been prohibited by countless laws."[2] In that particular case the Old Court denounced the use of a new piece of gear devised for gathering oysters, alleged to cause "damage to channels and destruction of all kinds of fin fish," especially among the underwater algae *[barri]* among the mud flats and swamps, "where the fish retire to feed."[3]

For the protection of small fry the authorities were especially unbending, and even their proclamations show their anger at a practice that in their eyes was absurd.

> Those evil parties therefore deserve great and severe punishment, as they completely ignore the law as well as the general welfare of this city; they are so wicked and cruel as to take a very small quantity of fry, from which they derive a negligible profit, destroying thereby huge numbers of fish which, if they matured, would provide the greatest abundance to the city, and great gains to those same lawless fishermen.[4]

Seeing such recriminations from the magistrates, surprising in their good sense and application to the present day, we must not fall victim to facile anachronism, a sin for which historians are never forgiven, and rightly so. It's not the politicians of 16th-century Venice who are anticipating our time with their old-regime economic rationality. It's rather our time, with its economy annihilating natural resources, that lets us rediscover the rationality of a world labeled "pre-modern" and "backward" in comparison with the triumphs of today's capitalist society.

The court's recriminations were not dictated by a desire to educate. They had to intervene day after day in a restless social reality solicited by shortcuts and frauds. The necessity of restraining abuses "that cause the shortage of fish in the City today," denounced by the authorities in the late 16th century, conscious of the fact "that the shortage comes from having exterminated fish before they were born, causing damage and incalculable harm," led the magistrates to enact very strict prohibitions aimed at protecting even the earliest stages of fish reproduction. Such destructive and wasteful practices had to be stopped by intervention at the source. "Let no one dare to take black goby with hand lines at the time when they have laid their eggs, and whosoever in any manner shall catch goby in the months of April, May, June, and July," when the eggs might be harmed, "if convicted, shall be condemned to the galleys for two years, rowing in leg irons, and to payment of 25 ducats, half going to the one who reported the infraction and the other half to the Lords of the Old Court; if not fit for the galleys, he shall spend 5 years in prison, in irons."

A similar measure concerned the blotched picarel *[menole d'Istria]*, "in the season when they are mating and bearing eggs, especially in the months of May, June, July, and August, with the same penalty, which also applies to anyone who has any kind of arrangement with whoever catches them, or buys them, salted or otherwise."[5]

There are harsh sentences, demonstrating the authorities' unbending condemnation of crimes that combined two infractions that were intolerable to the governing philosophy of the Republic: economic waste, and harm to the public interest.

The effort to control fishing operations in the lagoon so as to protect and preserve the reproductive balance of the various species of fish, was a constant aim of the Venetian authorities. A magistrate of the Old Court could rightly boast, at the end of the 18th century, that "the wisdom of our Elders, with excellent and salutary warnings, always aimed to control the taking of fry, since the supply of that essential food depends in great part on the fry; distinctions were made among the various seasons, places, nets and devices."[6] For that reason, at a time when various incentives were being offered for importing from other countries—because of increased domestic demand, and probably also a decline in the local catch—the Senate decreed on 17 May 1781 "as an experiment hiring special guards with the sole duty of patrolling the whole inner zone *[interno Circondario]* of the lagoon in the season when small fry are abundant, and to surprise and arrest the lawless violators who catch the fry, jeopardizing the breed and causing ruinous and immeasurable devastation."[7]

This preoccupation on the part of the Venetian authorities with conservation and protection, essentially aimed at the growth and reproduction cycles of fish in the lagoon, was plainly not dictated by abstract notions of respect for nature and her fragile biological rhythms. Even though the protection was extended to those aspects of life in the lagoon, the purpose was explicitly one of pure economics: to ensure the greatest abundance of fish that a respect for the *rules of nature*—growth cycles and life seasons among the fauna—could guarantee. The issue was adapting the food needs of the population better and more rationally, through technical control of fisheries, to the biological rhythms of natural production and reproduction that best ensured the abundance of the commodity, and potentially its existence undiminished in the future.

As the 18th-century justices said in reference to the taking of small fry, protection was essential for "that Gift of providence that prepares for the common benefit an abundant product ever born anew."[8]

4. Hatcheries in the Lagoon: The Fish Pens

This philosophy dominated fisheries in the lagoon, with original technical devices and very long-lasting institutions. It has not yet been mentioned, but a well-known supplement to nomadic fishing was "pen" fishing, the ponds being located mostly in the middle and upper lagoon. They were areas of various shapes and sizes set apart, usually next to a mud flat, often artificially shaped by carefully-built levees. Schools of various species of fish from the Adriatic, fry as well as adults, entered them following the tide at the coming of spring. The fish (bream, mullet, *boseghe*, *volpine*, *verzelate*) stayed in the pens until fall—eels might stay in the lagoon for six to eight years—and then tried to return to the open sea to spawn. But the pens were enclosed in different ways—either completely enclosed or "open," that is with lower enclosures—using reeds or other arrangements *[grisolie*, *cogolere*, etc.*]* to allow full control of the fishes' movements, in effect trapping them. Taking advantage of the instinct for returning to the open sea, fishermen provided the seaward passage with mazes and led the fish into prepared traps. The pens were generally closed from July to November and then were left open to the free flow of the tides.[1]

Like nomadic fishing, pen fishing was always a carefully regulated activity, down to the assigning of pens to individual *vallesani*, which was done each year at the beginning of March by means of a public lottery. The pens were of course subject to greater control, especially—as we shall see—because of their influence on the balance of waters in the lagoon, and because clandestine fishermen often tended to duplicate them in remote places where it was possible to enclose channels and swamps. But they remained an exceptional model of a fishing economy and a wise adaptation of age-old patterns of fish life in the lagoon to the food needs of the population. Indeed, fisheries in the *valli* represented the "raising" of fish stock, following growth and seasonal developments, and care was taken to "replant" as well. Usually in the spring, the fry caught in swamps and elsewhere, especially bream, several kinds of mullet, etc., were dumped into the pens, where they continued their growth. Meanwhile, according to a late-18th-century source, "From

1 March each year until St. James's Day in July it is prohibited to fish anywhere in swamps, mud flats, salt flats, reeds, minor channels, and heads of channels *[Cime dei Canali]*."[2]

By virtue of this economic process, based on respect and support for the reproductive rhythms of the fish, in the Middle Ages Venice could already boast a high degree of predictability in the product, both in quantity and in kind, and so could guarantee deliveries to monasteries and aristocratic houses.[3] Moreover, this was the policy that for centuries had governed the *piscinae neptuniae* (saltwater ponds)—that Cassiodorus believed to have existed before Venice herself—and all the various kinds of fish raising in the ancient world, of which the Etruscans and the Romans have left prominent evidence and lasting traditions along the coasts of the peninsula from Venice to Comacchio and from Orbetello to Torre Astura.[4]

Hunting and fishing—both subject to seasonal limitations—were not the only activities in the wetlands, however: there were other resource-based economies as well, and for a long time they were typical of many marshy areas on the peninsula. Reeds, straw, fertilizer, and hay were further renewable products used in crafts, livestock breeding, and farming.[5]

The Venetians' particular economic skill regarding raw materials and the goods necessary for survival—taking "economy" to mean a production and distribution system that places the highest value on available commodities and resources and strives to keep them renewable—was not limited to the natural resources of the lagoon. Even things that were not part of the fragile amphibious habitat where the city stood, primary goods imported from the mainland of Veneto or farther away were used with similar care, considered precious products, hard won from a world poor in resources, constantly threatened by natural hardships and men's greed, and subject to the slow rhythms of commodity renewal.[6]

5. An Appetite for Wood

Like all states possessing a fleet, Venice was forced to consume an extraordinary amount of wood every year, especially choice varieties, for her large-scale shipbuilding industry. But her needs were not limited to satisfying this significant requirement, with its variations over time. The city consumed wood of every kind for the most disparate needs, both domestic and industrial, whereas being built on the water she of course had no wood at all. So it is not surprising that the government and the appropriate authorities were constantly intent on conserving the sources of supply and ensuring a fair and rational use of wood on the part of the population. The authorities tended to intervene, regulating even the use of firewood for domestic purposes year round and especially in winter. This commodity was also subject to abuses that could hinder its circulation or limit the citizens' ability to purchase it. Early in the 17th century, for example, a decision by the Firewood Commission [Collegio delle Legne] gives a significant glimpse of the traffic in firewood within the city and the care with which the trade was regulated:

> In the many deliberations made at various times by our predecessors concerning firewood, there plainly appears an affectionate desire that the large population of this city should be able to procure cut wood when they genuinely need it at the price of twenty-five *soldi* a cartload; now although in the past hard times have caused some dissatisfaction with this necessary supply, the authorities never acted to change the aforementioned price, but determined instead to diminish the length of the wood rather than increase its price, as was wisely done on 27 October 1573.[1]

The authorities proposed to reduce the length of the wood from the customary two feet to one and one-half feet. To this end, our source continues, the Overseers of Firewood *[Provveditori alle Legne]* strove constantly to have "the greatest possible quantity for the use of the whole City" brought in, to avoid shortages and price increases. However, the resourcefulness and the political good will of the authorities were not always adequate to resolving difficulties:

> Since human sagacity persuades men to follow their individual interest rather than the public good, it has found a way to cut all the large firewood in the forests that measures a foot and a half, leaving behind all the small wood scattered through the forests; before long it rots, destroying the seedlings of future trees that rise every day in utter desolation; they bring the wood into the City and sell it for three or four *lire* a cartload after obtaining a license [from the authorities] first under one name and then another.

So men's "sagacity" not only evaded the law and harmed the common interest, the "public good," but damaged the fragile mechanism of renewal—the forest—by compromising the survival of various species when they were just seedlings. For these reasons the Firewood Commission prohibited bringing into the city "any kind of firewood of any other length than as specified above, that has been or may be cut on the aforementioned islands and in all of Istria, under penalty for the owners of boats, canal boats, and barges found in violation, of losing the whole amount of firewood that was cut disobeying the above order and brought by them into the city, and paying a fine of fifty ducats, half of which shall go to whoever reported the infraction, and the other half to charities *[luochi pii]*, with no possibility of pardon or gift or reprieve."[2]

On the other hand, scarcity of the commodity as well as attempts to profit from it on the part of the various carriers and retailers required constant vigilance on the part of the magistrates, who strove to prevent monopolistic buying and to regulate both the locations and the amounts for selling. In one of the many frequent proclamations issued in this regard by the Commissioners, on 7 September 1743 "sellers of vegetables and fruit" and other authorized venders were prohibited from selling faggots to innkeepers, boatmen, and other tradesmen, requiring all sales to be made "in their own shops for the aid of the poor and not exceeding ten faggots per day for one family."[3]

The Republic's concern for directly and constantly seeing to supplying the public led the Senate to take from the towns of the Dominion—which as we shall see were responsible for their forests—the responsibility for overseeing the distribution of firewood. The government had clearly seen from long experience "the poor management of the forests kept by the Towns and others lacking that legitimate and necessary supervision" demanded by the Republic. For those reasons, a decree dated 5 July 1700 ordained "the complete independence of Fuel Softwoods in the forests, which remain under our authority (the Overseers of Forests), since they are essential for the support of the state, especially the Metropolis with its large population."[4]

But Venice had a constant and particular need for timber, not only because of the normal and at times growing needs of construction or the ordinary uses by craftsmen and households—just as in every society of the old regime—but also because of her peculiar maritime location. In construction, for example, wood was used not just for the foundations of individual houses but also to create the quays called *"fondamenta,"* the space between the canal and the front of a building. In addition, navigational markers within the lagoon, the various elements separating the channels, such as *bricole* (heavy pilings driven into the bottom) or *pajne* (slender, flexible poles), and all construction in the water required wood.[5]

Different and greater were the specifically maritime reasons for the consumption of forest species. The city defended herself against the sea's slow erosion and periodic storms, as we have noted, by raising dense ranks of palisades, made up chiefly of logs, mainly oak, that were not marked for shipbuilding. Logs were also used for building the breakwaters extending into the sea and lesser forms of protection. It is reckoned that at the end of the 16th century about 140,000 logs must have been used in those constructions and spread along the barrier islands of the lagoon.[6] Not only was the wood subject to occasional damage, theft, etc., but most of all to the constant wearing action of the sea. In general the logs lasted no more than five years before they became waterlogged or riddled by worms, and they had to be replaced as they were broken up or damaged. For that purpose the authorities had always required that a stock of logs be kept ready in case of need. In 1664, as recorded by the treasury lawyer Calcaneis, the stock numbered 10,000 logs, and it was replenished each year with great difficulty.

By that time the shortage of wood had become acute: many forests had been cut down and turned to cultivation; others had been cut down to meet the needs of the Arsenale, or devoured by indiscriminate private consumption. The situation facing the authorities, a few decades before the ultimate solution was found of building masonry breakwaters *[murazzi]*, looked truly desperate:

> Indispensable for maintaining the inlets and preserving the Lagoon, for defending the City sheltering behind over twenty miles of barrier islands, with daily repairs and necessary projects absorbing huge sums, and the forests of the State and some neighboring Princes having been cut down, with the onerous purchase of logs from the Kingdom of Naples, and uncertain of the future, [the situation] has commanded the closest attention of this Magistrate.[7]

His lament came after a process four or five centuries long during which Venice had seen her consumption of wood for commercial and military purposes increase along with her rise as a Mediterranean maritime power. To supply the Arsenale—that gigantic construction site which in the 16th century was to become perhaps the largest industrial complex in Christendom—it had been necessary to secure huge quantities of high-priced woods, drawing from a constantly expanding area of managed forests. The "heart of the Venetian State"—as the Arsenale was considered—was the scene of a gigantic daily consumption of wood. Let us remember that in the early 15th century, for example, building a heavy galley required 380 bent oak planks, 150 straight planks, and 280 oak boards for the hull; 35 long larch beams and 18 larch or pine beams for the deck, plus 300 fir boards and other woods in lesser quantities, as well as hemp, pitch, etc.[8] We can see what the city's annual consumption of trees must have been after the battle of Lepanto in 1571, when Venice had to double the size of her fleet and bring the number of reserve galleys to 100, plus other vessels to be kept constantly ready in the Arsenale.[9]

6. Forest Management and Regrowth

In the second half of the 15th century, government intervention in the forest economy became more forceful and systematic. Until then the city had occasionally supplied herself with wood either by buying abroad or by drawing from the town forests—"crown" or public forests, dating from the [Holy Roman] Empire—bordering the lagoon and along the shores affected by it. After conquering Istria and Dalmatia in 1150, Venice gradually monopolized their forest production; then after 1340, following its purchase, the March of Treviso. Here Venice could use a forest, Il Montello, for centuries considered a jewel because of the quality of its oaks. Later the city had access to the forests of Padova after "annexing" it in 1405; to those of Friuli, Carnia, Cadore, and Belluno from 1420 on; and finally, after 1482, to the lands of the lower Po valley *[Polesine]* and the Alps. Growing control of the mainland led Venice to take a more complete view of her heterogeneous territory, considering her land resources as more intimately related to the economies and the needs of the lagoon than she had in the past. In July 1470 a law concerning the supply of oak *[provisio quercuum]* reserved all the oaks in the entire territory of the Venetian Dominion, wherever they stood, for shipbuilding in the Arsenale. This law not only transferred every oak tree, whether public or private property, to the state without compensation, but it also stated that the ground where it grew was perpetually bound to the growth of oak trees.[1] This severity was perfectly consistent with the shortage facing La Serenissima, but it could be adjusted in individual cases: for example, the inhabitants of Cadore, who lived exclusively from their trade in firewood, were exempt. It was also inspired by an effort to end unproductive uses of such a scarce and vital commodity. As recalled in the preamble to a law dated 24 September 1488:

> How necessary oak lumber is to our city everyone must understand, and since in the past we have been ruined for failure to anticipate future needs, and if we do not make provision in the present, such calamity will result that in order to meet the needs of our Arsenale and our whole territory we shall have to attempt at great expense to secure lumber from foreign countries, which will be no small danger for our state.[2]

The crucial political role played by lumber for the fleet led Venice to adopt a truly exceptional policy for valuable species. Whoever owned the land on which they grew, oaks were subject to special regulations under the control of the Council of Ten and the authorities of the Arsenale, and could be used for no other purpose. Individual trees bore the seal of St. Mark and were listed in special surveys drawn up at prescribed intervals. The trees were checked each year, and some were trained into prescribed shapes and bends, so that as they grew they would take on the profile required for a particular part of the ship under construction.[3] As for the forest of Il Montello, under the almost uninterrupted control of the Council of Ten and with a special Overseer from 1587 on, the Venetian authorities specified in 1668 that no nobleman could be elected to the post of Overseer if he owned property in any of the thirteen towns around the Forest.[4]

On the other hand, control over more common species was less rigid, perhaps in line with the broad autonomy granted by Venice to the mainland towns. In any case, policy regarding trees, both common and destined for the Arsenale or private shipbuilders, was very soon oriented towards economic rationality aimed at the renewal of the commodities being consumed. The specialists constantly advised and verified cutting and transporting "without injury to the young trees," and clearing the cuts of brush so as not to do "noticeable damage to the forest itself," as a chief at the Arsenale wrote on 12 August 1638.[5] The foresters of Venice and the mainland were already pursuing an active program of replanting certain species that was gradually transforming forest use into genuine cultivation, with human intervention cooperating ever more directly with the productivity of nature.

Forest cultivation lasted for a long time and was steadily refined. At the end of the 18th century that heritage had not been lost. A good example is part of the state forest of Il Conseglio in the province of Belluno, under contract to a private individual for cutting its plentiful beeches, the intent being to plant it with larches and other species for the Arsenale. The specialist who inspected the site after the cut advised the Overseers not to count on the spontaneous regrowth of the forest, according to opinions then current. The trees, he maintained, "might be thinned and lost because of the wind, and even if they were able to withstand it, they would suffer the further harmful effect of the spontaneous regrowth of beeches, which given their huge numbers in that forest may be considered entirely useless trees, occupying the ground unprofitably."[6]

Therefore, the Overseers had to plan a large-scale targeted intervention to program the growth of the desired new species—in this case mainly larches—imposing it on the mechanisms of spontaneous regrowth in a concerted technical effort. Renewal did not always follow a linear development in the automatic reproduction of commodities. Nature did not necessarily work for man in every case: she could also behave in most disorderly and self-destructive ways, as sometimes happened in certain forests. The chief for the forest of Il Conseglio complained in 1790, "Some trees in certain locations grow to be almost impenetrable and thus in danger of dying for lack of nourishment, since it is surely not possible that a piece of land that can only support a thousand trees may later support ten thousand."[7]

Long experience and observation had led to an "economic wisdom" made up of specific norms and knowledge, quite aware of the limits of both nature and human activity. It was no accident that the processes of chaotic disorder were replaced by a sort of rationality, a human attempt to achieve a richer, fuller "economy of nature." As recorded in a decree from the Arsenale dated 5 July 1709, concerning two stands of oaks in Treviso to be replanted:

> The young trees must be thinned, lest they not receive their proper nourishment from the earth or grow adequately, or receive less benefit from the Sun and not flourish because they were too dense; from the clearings of the excess trees, with the assistance of Chiefs and Experts assigned for this purpose, they must be transplanted to whatever parts of these two forests need them, and there not being enough seedlings for transplanting to cultivate the entire forest, many acorns must be sown in addition, so that every part of these woods may germinate.[8]

7. Forests and the Lagoon

In the forests of La Serenissima, then, not only were the growth cycles of the various species respected, but reforestation was increasingly practiced, a planned cultivation that was achieved by sowing, planting, and periodic cutting. These were techniques that required careful maintenance of the space that the trees needed for growth.

This practice deserves particular attention. Even within this specific natural habitat, far removed from the amphibious regions of the lagoon and the practices and cultural traditions of a coastal society, the Venetians tended to impose economic rules that respected and valued balanced productivity in the natural environment. Even the language of the authorities sometimes reflected the same concern for protection of the nascent forest commodities as was shown for the small fry in the lagoon. This is plain in the continual alarms voiced at the devastation caused not only by humans, but by animals as well: "Saplings and forests are destroyed in great part by the grazing of livestock, banned from the forests themselves, especially goats, whose fatal bite kills trees and leads to the total slaughter of the shoots."[1]

We must recall at this point that the authorities did not always reach their specific and general goals in forest management. Local communities and individuals, ignoring the reserve rules, often engaged in the clandestine destruction and cutting of timber. Not every town was disposed to submit to the harsh requirements entailed by the restriction laws. In this regard, forests subject to expropriation and managed directly by the state authorities were much better cared for and used more rationally. However we judge the final outcome of the Republic's forest management strategy, one basic cultural and political fact must be emphasized: even for resources located outside the area of the lagoon, Venice tended to impose general regulations aimed at protecting and improving yield, preserving natural productivity, and ensuring a steady and optimal stream of commodities.

We must also remember that the conservation restrictions applied by La Serenissima were explicitly meant to protect the whole territory, on which the health of the lagoon also depended. So it is not surprising that Venice began devoting special attention to the forests once she realized their function in preserving balance in the waters and on the land, and thus their significant economic and strategic role. One of the first protection laws we know, going back to 14 May 1282, explicitly [in Latin] prohibited slashing, setting fires and burning pine groves: "Let no one be allowed to cut or dare cut into any pine in a grove, or set fire to said grove *[aliqua persona non possit nec audet incidere aliquem pinum de pigneda, nec ponere ignem in ipsa pigneda]*."[2]

Most of all, starting in the 16th century—when the city as well as the entire Dominion was undergoing unprecedented population pressures—Venice's laws show the fullest awareness of the harmful connection between deforestation in the high country and altering the course of rivers on the one hand, and disturbances in the fragile habitat of the lagoon on the other. It's no accident that Sabbadino's technical study of problems in the lagoon sees in the destruction of the forests a historic man-made imbalance: "First, changing the rivers; second, destroying the forests; third, farming the mountains; fourth, filling the valleys."[3]

In that phase the human impact on the forests must have been particularly severe, if it's true that even private individuals began appealing to the authorities to stop the recurring destruction. "What's the use," asked a woodlot owner in Belluno, "of all the river diversions and other remedies and all the digging, if the same disorder goes on?" They are surely not a solution for the "great floods and the enormous quantities of silt and mud now being carried into the lagoon and deposited by the torrents and rivers."[4]

In the mountain hinterland of Belluno the growth of production (quarries, forges, etc.) drove the population and producers to indiscriminate plowing and burning. In the town of Agordo, a citizen complained in an appeal to the Firewood Commission, on 5 February 1698, that fires were a daily occurrence: "Now all the farmland and woodland has been burned and made to resemble the summit of Vesuvius, and there's not the top of a tree or grass or roots to be seen in any season of the year."[5]

So it was not easy for the Republic to defend the hydrographic balance of the lagoon on territory that was so far removed, while securing a raw material essential to her many different needs.

The disproportion between expanding needs and available resources did not allow Venice, any more than other great maritime powers of the time, to meet the need for wood by just managing the regrowth of trees in territories under her control. Resort to purchases from the Apennines and the Kingdom of Naples became essential in the 17th century to supplement the yield of forests in the Dominion. On the one hand, this was a sign of inadequate raw materials, linked to exhaustion of resources, and on the other of agriculture expanding into forestlands: the majority of the Venetian ruling class was not utterly opposed to this agricultural expansion, as it had been in the past.

Furthermore, as we shall see, in the mid-18th century the city successfully overcame the worst problems in protecting the barrier islands from the action of the sea. Barriers of masonry, the *murazzi*, replaced the log palisades with greater effectiveness, freeing the city from the burden of maintaining a huge supply of timber that had been a concern and a financial drain for centuries. Venice's decline as a maritime power starting in the same century put a permanent end to the contest. A city of stone, whose destiny was now of necessity turning landward, and ever more isolated from the new world economies of the time, she no longer needed to manage the renewal of her forest assets in order to preserve her inland waters and control the Mediterranean.

III

Technology, Institutions, and Regulations

1. A Peculiar Political Arena

Venice's location on constantly shifting ground, subject to repeated changes of various kinds, soon forced the city's ruling classes to adopt advanced and specialized skills. From the movements of the tides every six hours to the changing currents, from the wearing action of the sea to periodic silting up, everything in the lagoon required observation and adaptation, a shifting reality to which not only everyday life but government action as well had to be adjusted. Action brought knowledge of the complex movement of water within the lagoon and its rules, an empirical knowledge that became scientific hydrography, controlling the peculiar medium of community life in the city.

As we have seen, since the lagoon was also the site of economic activities, it required even greater development of technology, applying methods for using resources and obeying the limitations of an exceptional environment. Along with the technical efforts constantly demanded of men working on her inland waters, Venice had demonstrated another special skill by the late Middle Ages: the ability to create appropriate institutions to provide management and specialized knowledge for the most varied fields of activity. The government's political creativity in meeting the various and changing problems constantly brought forth new entities possessing the freedom to act.

The various authorities devised by the Venetians over the centuries represented in many ways the continuation of technical action by political means, crystallizing it in juridical regulations and consistent, uniform government standards. Practical modifications to the water and its many forms, and government by men responsible for meeting economic and moral standards of behavior, both converged in the various authorities: specialized entities for control and direction, characterized in their workings by collegiality and limited terms, which is a distinctive and very modern organizational feature of the Venetian State. The institution of the "junta" *[zonta]*,[1] a group of experts called in to assist an individual authority regarding particular problems, shows the constant concern of the city's ruling classes for combining government with technical knowledge, interventions with competence, and decisions with predictable results.

We should emphasize in this connection another reason for Venice's unique political and governing experience. It lay in the peculiar spatial relationship between the government and the governed. Unlike Europe's great nation states, the Venetian State was not the capital of a vast territory with a large and scattered population. Similar to what happened in the other towns of central and northern Italy, there was a closer, physical rapport between the lawmakers and those who were to obey and enforce the laws. But unlike Milan or Bologna, Venice did not stand on stable territory that remained unchanged over time: she lived on the water in a fragile ecosystem that was constantly changing, and she watched it day by day. Therefore the testing and evaluating of decisions made by public entities was carried out sooner and more constantly than was possible in any other state. The effectiveness of an institution or a new regulation could be immediately observed by the public, and thus examined and judged. Hence also the peculiar sensitivity of the authorities, their prolific legislation, their constant output of regulations, rules, controls, and revisions. In the lagoon, the fact that law enforcement could be promptly verified locally made for a unique relationship between the government and the citizens.

Using these tools, Venice faced and often won out over her problems, coping with periodic natural threats, uncontrollable processes in her territory, and the needs and pressures of men and their individual economic interests, which often conflicted with the long-range perspective of the public authorities, for whom the survival of the city was the ultimate goal.

Of course, the first combined technical and political undertaking of the city authorities was protection from the effects of the sea. As already mentioned, by 1275 two Superintendents of Barrier Islands were elected, and a third was added in 1281. How broad a field came under their control can be deduced from the numerous prohibitions they issued in the 13th and 14th centuries, outlawing the burning or cutting of pines, gathering oysters along the breakwaters, driving animals along the levees, uprooting reeds, etc. At the same time they required planting tamarisks, hardy amid salt and wind, and able at least to lessen constant action of the latter.

For a long time physical defenses against the encroachment of the sea were entrusted to many different structures subject to constant maintenance and periodic replacement, ranging from bundles of reeds placed at the less exposed parts to oak palisades. The latter could have one, two or three rows of pilings, usually reinforced by iron rods and bound together with strips of larch [bark] to form caissons: these were filled with rocks and reeds, and compacted with sand.[2] The wooden structures served not only to withstand the force of the sea but also as supports for channel markers at the seaward passes:

> Special care was taken, as stated by a specialist at the beginning of our century, in the building and maintaining of breakwaters or jetties, which extended out from the shore and thereby narrowed the passes. They were called "guardian breakwaters," as they guarded the inlets that feed the lagoon.[3]

That artificial line of defense quite effectively restrained the force of the sea as it drove towards the lagoon, although it was powerless against the great storms that raged for days and caused damage that could be seen long afterward: "breaks and gaps like the ones in the year 1600," the Sages lamented, "which cost 100 thousand ducats and did not lessen the damage to the lagoon, which will always show the effects."[4] But apart from their relative fragility, the real weakness of the defenses was the exorbitant cost of maintenance and periodic replacement. It was truly a "file gnawing away at the treasury," as the expression had it, costing the state thousands of ducats every year, like an endless, wasteful labor of Sisyphus. So all through the centuries of the modern era, specialists, experts, and individual writers debated and made projects, trying to find a solution that was in some measure permanent.

The solution came; it was late indeed, but it could rightly be considered permanent. It was an innovation compared to previous construction: hydraulic cement *[pozzolana]*, a material long unknown to the Venetians. The Romans had used it at Baiae on the Bay of Naples and at Messina, as well as in the ports of Civitavecchia and Leghorn. It was in Tuscany that Superintendent for Waters and historian of the lagoon Bernardo Zendrini encountered it, and he tried using it in the sea at Venice. A first experiment was carried out near "Malomocchio" *[sic]*, and it lived up to the hopes placed in it. The following year a breakwater was built, 16 paces long and 12 feet wide at the base, and it demonstrated conclusively the extraordinary strength of *pozzolana:* "The bond of this cement," wrote the Magistrate for Waters, "is so strong that it withstands every attempt at breaking it, and even more remarkably, the mix hardens under water in such a way as to become like stone in a very short time."[5] And so began the construction of the murazzi, the breakwaters of "Istria marble" and stones bonded with *pozzolana.* The project was authorized by the Senate on 4 August 1740 and furthered by decrees during the 1750s authorizing the construction of eighty rods [1,300 feet] a year of a long, artificial reef located in the zones that were most vulnerable to the destructive action of the sea. The operation was pursued for thirty years and only ended with the fall of the Republic; it cost about 20 million Venetian *lire.* It still exists today and was recently reinforced for the protection of the city. The masonry breakwaters had freed Venice from the crushing cost of the palisades, and in the period of the city's economic and political decline, they were almost her revenge on the waters that had always seemed to threaten her future.

2. An Authority for Water

From the 14th century on, at least, the Venetian authorities saw the need for instituting specific authorities charged with overseeing the waters of the lagoon. In 1324, Zendrini records, four Sages were elected to oversee the swamps, charged with finding remedies for the damage, by now visible, caused by fresh river waters flowing into the lagoon. "This was the first time," he tells us, "that the Republic considered separating the fresh waters from the salt, chiefly where they directly threatened the illustrious city with silting up and spreading of the swamps."[1] But they must have been unstable institutions and were probably ineffective, because they were followed by an endless series of other authorities and new initiatives. In 1399, Cessi records, a Commission of Sages was instituted to undertake a far-reaching rehabilitation of the lagoon. The results must not have been outstanding, because the overall problem of the lagoon was taken up not long after, entailing a new commission composed of seven Sages. In 1415 six more Sages were added to the first seven, without appreciable results.[2] Meanwhile, however, the city's power had extended beyond the limits of the Duchy, first with the acquisition of the March of Treviso in 1338, then during the 15th century reaching the greatest extent of her mainland dominion. The Republic's control over greater territory soon was reflected in a much broader technical awareness of the lagoon and its complex problems, along with new intervention policies and new authorities with a broader strategic orientation.

Indeed, starting in the 16th century a definite acceleration is seen in legislative response, with the creation of new authorities in every area of Venetian life. The conquest of the hinterland, the mature power of Venice within the Mediterranean economic world, the extraordinary pressure of population on existing resources, and the emergence of a real trend towards farming in many sectors of society, all these led to the creation of more stable institutions capable of presiding over large-scale improvement projects to be pursued over time. Significantly, the century began with an initiative destined to become a fixture of rational and farsighted policy for the lagoon. In 1501 the Council of Ten created the Authority for Waters *[Magistrato alle Acque]* by electing three Sages for two-year terms; they could enter the Senate chamber (but not vote) when issues concerning the waters were being discussed. In this way all the previous authorities, like the Town Overseers, were incorporated into the new institution, which was charged with verifying the condition of the levees in the lagoon, reviewing leases, "securing the removal of all buildings and structures on state-administered land," and seeing that "any and all were prohibited from filling even the smallest portion of the public waters, with a fine of one thousand ducats. Finally, supervision was ordered of everything concerning the health, the security, and the convenience of the city with regard to the lagoons surrounding it."[3]

They had the additional burden, the "express obligation," as Calcaneis recorded, "to have all the levees surrounding this lagoon examined every year by the Chief, the Vice-Chief, and their assistants." The procedure was carried out in July and was followed by a "sworn report of the condition of the levees, of their deficiencies and their locations, for making the appropriate corrective remedies."[4]

In 1505 there was discussion of a High Commission for Waters to strengthen the new authority, which would meet in specific cases for special deliberations. At first it was made up of fifteen Senators, but later its numbers varied, reaching seventy-five members. They included the three regular Sages for Water, forty Senators, three Executives, the State Attorneys, all the members of the Full Commission with the Doge presiding, etc. As Abbé Tentori records, one of the special prerogatives of the Commission was electing the members who from time to time were called on to lead it, a testament to the power assigned to the authority and the relative independence of action it enjoyed among Venetian institutions. It met at least once a week, and in 1587 the Doge was required to convene it periodically.[5]

This new authority consolidated its power especially after 1520, when it took over supervision of the barrier islands as well, till then largely the province of the Salt Authority *[Magistrato al Sale]*; over the centuries it became one of the central institutions of the Venetian State. The Commission was given the power to levy customs duties, and by Senate decree of 1565, the right to impose a tax of 5 percent on the capital of every inheritance, which the heirs had to pay to the treasury when they inherited. This levy was called by Venetians "the water payment," and over time it represented a significant source of revenue for the Commission's operations and purchases. Surely the ultimate seal of dignity—in addition to the presence of an Inquisitor with the right to hold trials—was an action by the Council of Ten in 1505 granting the Commission such great independence and deliberative autonomy "that membership (like membership on the High Commission) was closed to all Nobles who owned property or any other thing on the Venetian lagoon, rightly considering that they might place it ahead of the general good,"[6] evidence of the predominance in Venice of public ethics, a point to which we shall return.

If the Water Commission had technical duties great and small to discharge, they were almost never neutral socially or exempt from conflict with the productive classes whose activities took place in the lagoon. We have already seen how strictly fishing in the lagoon was regulated and how harsh the penalties were for infraction. But we shall not fully understand the eminently public content of the restrictions until we consider an aspect that has not yet been emphasized. In Venice, fishermen were a class unlike any other. They were constantly out on the lagoon, "plying its waters day and night,"[7] so La Serenissima gave them the assignment of keeping watch, observing the changes that took place from time to time and reporting significant alterations.

Men, families, and groups who lived on the water and lived from it were also called on to perform a significant public service by informing, reporting, furnishing technical data, and even by recalling changes observed in the past, sometimes in the course of a lifetime, in the various parts of the lagoon. They were the constant, living record of the health of the lagoon, which in a sense stayed alive because those men worked there. La Serenissima also acknowledged the fishermen's economic value with public recognition: "Such a deserving class of people," read a proclamation of 19 January 1780, "who risk their lives and their fortunes to promote abundance for the public benefit, shall be preserved in the peace, security and tranquility they deserve, and protected in their just interests."[8]

And yet, even with regard to persons considered so deserving, the force of the law was applied with impartial severity when any of them, "disregarding the common good for his own," violated the rules established for fishing: using prohibited nets, taking fish during the spawning months, enclosing open waters. Then heavy sanctions took effect, ranging from impoundment of the boats and nets—they were usually burned—to monetary fines, the galleys, or a term of years in prison.

The ultimate aim of conservation efforts on the lagoon could come into open conflict with one of the oldest fishing practices: the pens. As had been observed since 1314, and as Sabbadino convincingly showed in the 16th century, by using reeds or woven fencing to build enclosures in flowing waters, and by taking and using mud that often wound up in the channels, the pens fostered imbalance in the waters of the lagoon.[9] The government intervened several times in 1536 and 1559 to remove from the lagoon individual enclosures judged to be harmful. So in spite of constant vigilance, and occasional instructions to break open enclosed pens in cases of obvious degradation of the overall water condition, especially in the Malamocco channel, in 1662 it was decided to destroy fully fifteen of the twenty-eight pens in existence, located in the middle lagoon; they were "returned to nomadic fishing only."[10]

3. The Rule of Law

It was not the fishermen, however, who required the greatest effort from the Venetian authorities to enforce standards. In the mid-16th century the Republic was in the midst of profound economic and social changes that brought new duties and new challenges. Population growth and the influx of peasants to the city were expanding the demand for agricultural commodities, there was increasing determination to end dependence on imports from southern Italy and Turkey, new and broader sources of revenue required taxation, and investment capital was moving from commerce towards real estate. All these factors were turning Venice towards the mainland and agriculture. We know that eventually the push to improve the land, the reclamation and draining of ground in the Dominion, came into conflict with the effort to preserve the lagoon. The dilemma is exemplified in the conflict between Cristofaro Sabbadino, a zealous guardian of the inland waters, and Alvise Cornaro, a champion of land reclamation and expansion of the agrarian economy.[1] In spite of disputes and controversies, the Venetian authorities responded to the new needs with a steady and consistent policy of regulation. On 19 September 1545 three nobles were elected with the title of Overseers of Undeveloped Holdings [*Provveditori sopra i beni inculti*], and charged with promoting and regulating land reclamation and irrigation on the mainland. Thereupon consortiums sprang up and spread, organized by farmers in a collective effort to open and drain new lands or else to organize irrigation systems. Private initiative had to adapt to a dense network of regulations (and inspections) in order to pursue its goals:

> To allow the cultivation of arid or unworked lands, the Leader or Consortium intending to take water from rivers or irrigation canals or other underground sources shall stake out the entire course of the planned canal and make a map of the land intended for irrigation, as well as the canal from its intake point to its outlet, and submit it to our Overseers of Undeveloped Holdings, who shall send to the site two experienced Experts of their choosing, at the expense of the Leaders.[2]

It is useful to recall in this regard that on 22 April 1527, in order to keep "public service" separate and distinct from private interest, it was debated whether to appoint three members of the Senate for the purpose of replacing Overseers in cases where the latter "had property, or a father, children, brothers, or sisters, parents-in-law, sons- or daughters-in-law, cousins or brothers- or sisters-in-law . . . involved in the Consortium being created."[3]

Of course these and other rules, like the requirement that anyone intending to "make canals" between the Brenta and the Piave inform the Sages for Water because of the immediate impact they could have on the waters of the lagoon, or the requirement that land between the river mouths and the shores of the lagoon be two percent reforested by their owners, were not enough to restrain the tumultuous push of private interests to expand farming to areas reserved for what we might call the "public economy of the lagoon." But we must remember how the Republic tried to contain economic initiatives within a framework of cooperation among the parties, even outside the lagoon habitat itself. The experts always tried to impose on the consortia, even in cases where "fish-pen" land was being drained, the principle that equal and proportional contribution was "just and appropriate; therefore, those who use and benefit from said Consortium . . . must also share in the expenses already incurred, as well as future expenses."[4] It was also required that private gain not be transformed into public harm, as could result from converting dry lands or swamps into rice paddies. So the conversion had to be authorized by an expert opinion, after a visit to the site. "As to the harm that the water might inflict on hamlets in the vicinity," said one expert in 1582, "I affirm that . . . the following regulation will prevent harm to any neighbors, whether mills or buildings of any kind whatsoever because the water, being shallow, cannot flood any land or jeopardize buildings, because there are none of any kind there."[5]

Experts were always present, even in cases representing private individuals, who—especially in the province of Verona—usually presented requests for the use of water for irrigation. The typical request, like that of a Verona landowner in 1577, was for "trying, with what little digging I can afford, to find a little water." As required by law, the request included a detailed drawing of the area.[6] Approval of the request for land modification was always conditioned by respect for neighboring interests. "We have ruled," experts declared concerning a Verona landowner's request "near Il Chievo," "that said water may be used by said applicant," but with definite restrictions. The experts stipulated that the projected canal must flow within well-defined limits. They were convinced, "to prevent damage to mills in the neighboring hamlets, that the path of said canal marked in red on the drawing must conform to the course laid out in the experts' report."[7]

So by drawing "red lines" on maps of the territory, Venice's authorities were trying to apply on the mainland the rules of social conduct embodied in policy for the lagoon: private interest in harmony with the public interest; individual use of the land's resources without harm to others. Of course it was not easy to reach such a lofty goal, especially among populations who felt dominated by La Serenissima rather than governed by her. They only felt the negative effects of conservation in the lagoon, like diverting major waterways from their natural outlets. We must never forget that consensus was always the firmest foundation for the ruling classes in their efforts to protect the lagoon.

The lagoon could not always be protected from interference in natural habitats so far from its shores. Especially in the 16th and 17th centuries many of the changes felt in the inland waters resulted from actions undertaken by private individuals in remote places, in the mountains and the forests. Already in the 15th century the authorities were fully aware that "said deforestation is an obvious cause of silting in our lagoon, for the rains and other waters are not restrained or stopped by the forests from pouring into the lagoon, as they once were."[8] Within the boundaries of the lagoon, the clandestine filling of swamps and salt flats, the building of levees, and the expansion of farming and grazing must have increased, judging by how often they were recorded by the sources (and the writings that illustrate them). The Commission for Waters, in one of its many measures, blamed "the poor condition" of the lagoon

> principally on the many infractions and encroachments made despite the provisions of our laws, especially the one dated the last day of April 1562 and read today, such acts being committed by various Nobles, Citizens and others. Not only have they failed to respect their country as they were bound to do above all things, the preservation of which ought to be the aim of every good citizen: they have not obeyed the laws in any measure, but have again enclosed many areas of open water, openly encroaching on the public salt flats, and digging many drainage canals and connecting public channels, planting and sowing just as if they were on the Mainland."[9]

The penalties for such infractions varied, especially for repeat offenders, including restoration of the site at their expense, confiscation of the land (to be shared among the enforcement officials, the person reporting the violation, and the water authority), destruction of trees, confiscation of animals, etc.[10] Following principles that are only in part foreign to the judicial mind today, the Venetian authorities specified in each law the advantages and benefits guaranteed any citizen who reported an infraction. This provision was surely not without drawbacks and risks, even concerning the proper enforcement of the law, but concerning damage inflicted on the territory it gave the magistrates better opportunities for verifying the infraction and was some protection against errors and arbitrary judgments. Moreover, since the Republic had to watch over such a large territory, only the cooperation of all citizens, motivated by self-interest, and the denunciation of violators by third parties, could ensure greater awareness of interventions and damage, while it discouraged those who might have counted on unreliable or hesitant accomplices. So it is not surprising to find among the various measures taken by the authorities an exhortation to citizens who discover enclosures and various kinds of illegal fishing in the lagoon to "report the location of the clandestine enclosures, and receive the entire monetary penalty; all the boats found in the enclosure shall be sold, and the money given to him secretly."[11]

So the Republic found itself fighting hard against the economic forces that sought direct and unlimited access to the lagoon and its resources and commodities. Whoever wanted access to the waters and the economic opportunities they offered had to obey the public regulations issued and renewed from time to time: these represented the government's intention to preserve the hydrographic and environmental balance of a fragile urban setting and allow the existing natural commodities to reproduce and renew themselves according to compatible biological rhythms. Here we must note what is surely one of the most original and modern aspects of Venetian governmental behavior. In order to preserve the lagoon—in a word, to guarantee the conditions for Venice's very existence—every private interest, however great or significant, had to be subordinated and adapted to that ultimate, unquestioned principle. A higher outlook, reflected even in daily behavior, was imposed as an egalitarian norm on all citizens and all classes, impacting and shaping the very structure of the State. The necessity of controlling and protecting the unique and fragile territory on which it stood, with the technical and financial investment it entailed and the complex network of institutions and laws, had become at a certain point the very substance of the State. As declared in the decree dated 18 May 1505, establishing the High Commission for Water, "the issue of water is of such weight and moment that it may be said to embody in a single word the substance of our entire State *(haec materia aquarum est tanti ponderis, atque momenti, ut unico verbo dici possit importasse secum consistentiam totius Status nostri)*."[12]

It can easily be imagined that such a harsh system of restrictions was not without drawbacks and anxieties for productive interests, especially when signs of degradation in the lagoon sharpened the attention of the authorities. There was no lack of complaints and appeals, like the one dated 22 August 1538, in which a certain Bartolomeo Moresini and his brother requested inspection of a palisade built in a channel which they claimed as their property, "*nostro proprio*." He complained, "we cannot plow, build levees or drainage ditches, graze livestock, build or do anything that harms the lagoons, on eight parts of our property," adding that under those circumstances "we are deprived and despoiled of what is our own."[13]

This was, inevitably, the price to be paid when the location of private property and productive activities had immediate effects on the movement of the inland waters. In other cases it was the effects of productive activities that sparked controversy between private individuals and the authorities. Almost never were the repercussions of enclosures upon the flow and the circulation of the waters universally agreed upon. "I declare," a landowner explicitly maintained in 1544, "that in my opinion placing reed fences and making fish pens improves the bottom rather than damaging it." And as confirmation he recalled that "where there are pens the swamps never dry up."[14]

Let no one think that Venice employed a vexatious and totalitarian system of restrictions for the lagoon area. The government well knew it was dealing with economic interests that had a legitimate need for expression, and thus for using public spaces and resources. Nor were the authorities unaware that they needed to secure the *common interest* of private individuals in the non-destructive and, if possible, renewable use of the lagoon habitat.

4. Legality, Equality, and Liberty

Venice's ruling classes, trained in commerce requiring long voyages, investments, and risks, had a dynamic notion of private wealth and its value. On the other hand, preserving the lagoon's ecosystem was not without its cost. Simply respecting the status quo, as imposed on the citizens by the authorities, would not have been enough to save the city: on the contrary, a constant and concerted effort was required to combat natural tendencies that otherwise would have had a certain and disastrous outcome. And making that effort often required huge financial means. A continuous return of wealth through fiscal pressure was required from the various social classes, and mobilization of a consensus that the State never could have achieved through just a system of restrictions and prohibitions alone, unilateral interventions that worked *against* individual freedom of action. Such a course would have led to a despotic system, a "hydraulic monopoly" like the ancient world as described by Wittfogel in *Oriental Despotism*, destined to expose the State to self-destructive social conflicts. Venice was not a peasant society, universally subject to forced labor. Her strength was founded on private trade, on the exercise of free commercial activity. Laws and prohibitions, even when limited to a particular area like the lagoon, would have become intolerable if they had only been prohibitive and hostile to economic free enterprise.[1]

In fact, the Republic left broad economic and territorial areas—lots, fish pens, lakes—to private ownership, requesting landlords and tenants not to engage in productive activities incompatible with the preservation of the lagoon. In addition, with flexible political pragmatism the Republic herself constantly sought actively to involve private citizens in the onerous work of protecting and improving the lagoon habitat. The goal was pursued through a multitude of relationships and contracts often aimed at favoring the private use of public commodities, as for example the leasing of water and other areas of state interest. But the government often resorted to less traditional methods, for example, selling salt marshes to individuals or assigning them to whoever requested them for the purpose of turning them into fish pens. In this way the Venetian authorities achieved their goal of assigning to individual citizens the task of eliminating landfills in various sites and canals, without burdening the public treasury.[2]

We must not forget that Venetian citizens were expected to follow general rules, restrictions that affected everyone without exception, rather than improvised or arbitrary laws or shifting, whimsical measures. The shadow of arbitrariness was always kept far removed from the promulgation of laws and the severity of their enforcement. It was no accident that the Venetian authorities always bore in mind that officials charged with "enforcement of the laws might fail to perform their office or interfere in other matters, or going from sins of omission to those of commission, might commit extortion, assault, or oppression." Therefore they prohibited "said Ministers," when called on to oversee the conduct of fishermen in the lagoon, "from daring on any pretext whatever to accept Gifts, Donations, Payments on Account from Violators, by such connivance giving occasion and encouragement to law breaking." In this case, "their guilt was considered a crime against the public, an injury to the common interest."[3] How exceptionally modern these words sound, and how profound this knowledge of social realities that sometimes involved the public representatives of the law, the guardians of their enforcement.

Let us recall, as already pointed out, that Venice's government in the six or seven centuries of her glory was certainly not a democratic Republic like those now found in so many different societies in our time. It was characterized, at least till the end of the 13th century, by unusual forms of "popular" participation in the managing of public affairs, the *res publica.* But the exceptional modernity of Venice's governance that surprises and fascinates us today must not lead us into the old sin of anachronism. In reality the public life of La Serenissima was entirely in the hands of the nobility, who constituted a small elite. After the closing *[la Serrata]* in 1297 of the Great Council, which was the center of Venice's political power, participation in the political life of the city was limited to those who had taken part in it until then regardless of their ancestry, and of course their descendants. That body appointed or elected all the nobles who were to occupy posts in government, the courts and the administration. In the 16th century there were about 2,500 members in the Great Council.[4] But the limits on social representation in the courts and the governing bodies, the limited range of classes that managed power—as in almost all societies under the old regime—should not make us forget that Venice's public action was a genuine precursor in its universal scope. Scrupulous to the point of Byzantine complexity, throughout the centuries Venice sought modes of election to the various government posts that avoided such ills as creating factions, making power personal or hereditary, letting terms of office last too long, or letting family interests win out over the general public good.[5] These modes applied to everyone, and penalties could be imposed on anyone, without distinction. According to a recurrent phrase in the Code, every law concerned "everyone, Nobles and Citizens, and Churchmen as well as Laymen, excepting no one." "Excepting no one" was the dominant theme for obedience, because according to a motto of which the Republic was very proud, "Venice's affairs are governed by law."[6] The principle of equality before the law was so rigid and so strictly enforced as to impress such an outstanding observer as Jean Bodin, the French theoretician of monarchy, who wrote at the end of the 16th century: "An offense given by a Venetian gentleman to the least inhabitant of the city is corrected and punished with great severity, so that life for all is most sweet and free, more like popular liberty than government by aristocracy."[7]

Entrusted to a complex system of checks and balances in the various authorities—not excluding overlaps and conflicts—the government of La Serenissima always aimed to conduct public affairs on a plane higher than the material interests of the groups and classes it represented, although those interests were considerable if not dominant. In the delicate and ever-changing issue of the lagoon, government always proceeded experimentally—"for the sake of experiment" was the phrase—always ready to reexamine or change or turn back when faced with the results; always guided by the unquestioned requirement of preserving the inland waters, the ultimate goal to which all interests must bow, and towards which they must cooperate insofar as possible. Regarding the goal of common security, everyone—nobles and religious, rich and poor, powerful magistrates and simple laborers—was forced to put aside particular attributes and privileges and become members of a "community of equals," donning the modest sash of responsibility involved in being citizens of Venice. Only a State maintained by a spirit of higher equity and a goal of protecting the common interest could expect such behavior and such an attitude.

It is also interesting to note that in expecting citizens to respect the lagoon, the Venetian government not only expected consensus per se, acknowledgment of its own power and obedience to the rules that guaranteed the material order of its authority; it also expected recognition of a third agent, external to the sovereignty of the State, neither a power nor a divinity. It was simply the habitat of the lagoon: the essential condition for everyone's life and survival. Freely flowing water, navigation, and security represented a shared wealth that superseded all individual interests. So the citizens were not only expected to obey, but also to cooperate knowingly, behave responsibly, and demonstrate an extra measure of moral commitment. The symbol of their common destiny, their fragile and threatened habitat, embodied a principle of community that all Venetians were encouraged to recognize and share.

Beyond respect for the lagoon, the Venetian ruling classes were calling on citizens to recognize as well the superior justice and merit of those who were requiring and enforcing that respect. And so by protecting the general interest the authorities were on the one hand reaching the universality of modern statecraft and on the other hand making all the members of the Republic's community recognize that universality. Therefore, it is easy to see today that a considerable part of State sovereignty was grounded, so to speak, deep in the waters of the lagoon.

For this reason the cohesion of Venice's community structure—notwithstanding all the conflicts and clashes that shook it, like any other society on this earth—reveals the golden threads with which the daily defense of an absolute and indivisible wealth continually bound government and the governed together in a special relationship. In reality, on the waters of the lagoon the Venetians had built a powerful structure of consensus: the very one that today's governments, commanding the whole planet and faced with critical environmental problems, still seem far from seeking.

At this point a special meaning attaches to what Frederick C. Lane, a great historian of Venice, emphasizes concerning the myth created by the city, at least from the 16th century on: that its kind of power, its art of governing, was superior and unique.[8] The Republic tended to describe itself as a homogeneous and unified body that worked impartially and fairly in every sector of community life. Concerning its past history and its present conduct, it tried to create a halo of legendary, almost sacred superiority. But this effort, which should warn the historian not to confuse ideology with actual processes or contemporary self-image with the facts, is a precious piece of the whole puzzle for us. The continual praise for Venice's justice and good government so frequently found in the public records, so often found like a ritual at the start of even the most modest technical report, an insistent theme even in accusatory appeals made by individuals to the authorities, is nothing other than a mighty cultural effort on the part of the ruling groups to shape collective awareness among the citizens. By repeating that image of itself, making it a commonplace, so to speak, a part of current usage, the Republic was creating what today we would call "Venetian public spirit." Equity and justice in the conduct of public affairs, impartiality in dealing with citizens, the highest reverence for the lagoon in opposition to any other goal or interest, made up not only the credo to be followed by authorities and government officials, but the nucleus of a kind of secular religion that was to reach the furthest recesses of social awareness at every level: the governing instrument *[instrumentum Regni]* used in responding to the constant and diverse challenges made by Venice's physical location and political situation.

There may be no more eloquent or focused testimony to this ethic of public supremacy than the reflections of Abbé Tentori, written as the end of the Republic was drawing near. His considerations have a bearing on today's events. Since the first concern of the "Princes" (the governing body of Venice) "was always trade," he writes, "for this reason the control of it was never entrusted to Merchants, chiefly intent on their own private gain with no regard for the common interest, whereas the common benefit must have a General orientation, such as envisaged only by Governments."[9] Governing a country or running a State, as the Venetians well knew from the Middle Ages on, is something much broader and more complex than taking care of the economic interests of a group or a class.

Leading players for so many centuries in trade and commercial adventures on the world markets of the time, with men and assets in every corner of the world, the Venetians were more familiar than most with the process of exchange and economic competition. Therefore they well knew that without the higher mediation of the State, without the complex and farsighted art of governing, no providential "invisible hand" would ever restore harmony among the unruly actions of individuals.

IV

Banishing the Rivers

1. Daily Maintenance

The enemies of the lagoon, as Sabbadino recorded in the 16th century, were not just the sea and men with their selfish interests, but also the most powerful and unruly natural forces that invaded the inland waters: the rivers. As Marco Cornaro had already steadfastly maintained, the damage wrought to the lagoon by the action of the sea was nothing in comparison to "the harm done by raging rivers,"[1] that vast tangle of watercourses large and small, canals and millraces, flowing from the Alps and the territory of Veneto into the sea. Here was the source of the greatest and most constant threat to the lagoon, especially from the 15th century on, when increased deforestation and expanded agriculture sent rocks, mud and sand downstream: the deadliest of poisons for the inland waters. It was no coincidence that the Republic soon created special technical positions charged with the supervision of major rivers. "The River Expert *[Perito a' Fiumi]*," Tentori records, "usually called Chief Engineer *[Proto Ingegnere]*, was to receive detailed information on the rivers of the Venetian State, chiefly those which are navigable: their levels, the nature of their courses, and the greater or lesser volume of water they contain at various times."[2]

To combat the continual process of silting caused by rivers, Venice developed over time an elaborate strategy carried out on two different levels: almost daily emergency intervention in the lagoon, remedying the visible damage as it occurred and trying to keep it from growing worse; and large-scale projects carried out in the near hinterland to divert rivers and thus radically eliminate the sources of silting.

Constant soundings made from boats by the experts and regular dredging of the lagoon to remove sand and mud were ancient practices in Venice, and they answered several purposes, both short and long term.[3] As Alvise Zorzi, a Sage for Waters, recorded in 1589, one of the "most important and necessary projects that can be carried out now" to improve the flow of water in the lagoon "is to pursue most diligently the dredging of main and navigation channels" that had been resumed in the previous three months, "to facilitate navigation and the transit of merchandise." The purpose and effect, "besides improving the currents, preserving in this way the essential work done in recent years dredging the canals of the City, being to improve the flow of the incoming and outgoing tides, to the great benefit of the Lagoon and the Port"[4]

Making the inland waters navigable by continuous and systematic dredging not only favored the movement of goods, but it also contributed to safeguarding the spontaneous mechanisms whereby the lagoon tended to renew itself by mingling with sea water as a result of the tides. Material efficiency and healthfulness were inseparable as usual and had to be pursued simultaneously.

The Venetian authorities made use of both routine and special dredging. A law enacted by the Senate on 4 August 1565 levied new taxes to cover the expense:

> Recognizing that no remedy is safer or more effective than continual dredging in order to preserve the lagoon from filling in, without regard for expense or anything else, to avoid the grave and enormous harm that would result for the city from such a filling in, our College for Waters has resolved . . . that three thousand men be brought from the mainland to dig and scrape the flats.[5]

This was a practice that antedated the "Supervisors of Discharge" of mud [*soprastanti agli scarichi*], who were to direct the dredging and transport, taking care that "there be no risk of spreading it in the Lagoon or channels; therefore they are required to be present in person from the first daylight hour till the last evening hour."[6]

For a long period of Venice's history, the dredging of mud from the waters had been a condition for the very growth of the city. Here we may recall that her site and even her territorial stability were the fruit of a slow but powerful reclamation program. Constant work in draining, transporting material and building levees had been necessary for the citizens to build their land. As a 12th-century [Latin] source cited by Filiasi records graphically, "They drained swamps, made ground, and supported buildings almost on the deep *[siccaverunt paludes, manufecerunt solum et quasi ex abyssis aedifitia sustulerunt]*." The mud was used—as Tentori records in his work cited earlier—"for enlarging the Islands, forming new ones, and expanding the site of the Ruling City." And in this way the State tended to finance the dredging itself, either with transit duties to pay for work on the channels or with an additional tax on ground where silt was dumped.[7]

Especially until the 16th century, when the territorial profile of Venice and the islands became fairly complete, the city had fed on earth, expanding and consolidating the ground with material taken from the water. In the continual dredging operation intended to prevent the lagoon from filling in, the Venetians found value in the materials the rivers brought from the Veneto hinterland, turning a mortal danger into a means of reinforcing their amphibious settlements.[8]

This was a great and tireless effort on the part of the Venetian community, which worked conscientiously every day against the irresistible natural processes, trying to shape environmental conditions to meet their needs. For example, they often dredged new artificial channels to increase the flow of the waters and improve its circulation. These projects resulted in the so-called digs, the *Cava Zuccarina*, the *Osellin*, and the *Cavallin*.

A few cases also involved private individuals. At the end of the 16th century Marco Zuccarini, owner of the *Cava Zuccarina*, recalling that his ancestors owned it for ages under the name of *Cava d'Arco*, proudly informed the Sages for Water that they "always kept it in good repair and so well dredged that every kind of boat, whatever its size, goes through it with ease."[9] But dredging in the lagoon was work done "downstream" from the events that occasioned it, or as Contarini put it in the 17th century, "medicine for a disease already present." The large rivers emptying into the lagoon constantly added deposits of sand, and at that time in history the volume of deposit was greater and the rate faster, compared to the dredging.

In fact, as the specialists and the Venetian authorities were soon forced to recognize, problems related to large rivers had to be dealt with directly, before they reached the basin of the lagoon. The first move in that direction was one of the largest initiatives undertaken by Venice in the course of the 15th century. On 28 May 1412, it was decreed that "the banks of all the rivers in the State, and those of the lagoon, are declared to be under treasury authority and government administration, therefore they can in no way be sold or leased or leveled by anyone." From then on, whoever wished to lease land, water, or fish pens had to obtain a permit from a special authority whose members were recruited from among the chief offices of the State.[10]

With this decision the Republic succeeded in extending around the lagoon and beyond a process that had been initiated by the ruling classes in 1282 with the creation of the Planning Commission, an institution that oversaw the issuing of building permits, the siting of industries, and the condition of thoroughfares on land and water; it had taken on the specific task of recovering for the government many sites illicitly occupied by noble families or religious communities. The great work accomplished over a century by this institution, as documented in the precious Planning Code *(Codex Pubblicorum)*, had created the conditions for broad public control over a lagoon area that now extended to remote locations beyond the inland waters, but was intimately related to the balance within them.[11]

Once the whole territorial dynamic involving the lagoon became clear, public control of rivers was a perfectly natural measure, the more so because more and more often their normal flow was being tampered with by individuals who, according to a law of 1501 concerning damage to the levees on the New Brenta, "ignoring all public regard and intent on their private gain, take the liberty of breaking, leveling and cutting the Levees." In the hinterland as well, dredging the riverbeds to improve the flow and adapt the river to individual needs had become routine—sometimes legally requested of the authorities by individuals—to such an extent that besides "damage to the poor land," as another law put it in 1535, "it often causes them to capture the water, to the detriment of the lowlands and therefore of our lagoons." We must remember that the preoccupation on the part of Venetian authorities with preserving the beds and the banks of streams was not just aimed at safeguarding the narrow needs of the lagoon basin. The goal was much broader and more general: that of hydrographic balance within the hinterlands, as emphasized in a law of 1568: "The conservation of said Levees, and Ways most convenient for our affairs like River Navigation, is essential so that our Faithful [citizens] may enjoy in peace the fruits of their property, without any neighbor or other party causing floods or water in excess of that which they usually have." The conservation of levees was considered of such general value that whoever damaged them or the public ways "with injury to a third party, once the facts are proven, will without remission suffer capital punishment."[12]

Once again we must recall that in this phase agriculture was expanding, both on the shores of the lagoon and the rest of the mainland, chiefly through the use of irrigation. As a historian recently recalled, another significant factor was land reclamation, winning new lands for farming, in most cases involving increased use of water. Expanding and improving agriculture in those lands meant draining marshes or digging canals from rivers large and small. By the end of the 15th century a large river like the Piave, for example, was already the object of systematic (and, it seems, well-designed) projects for agricultural purposes.[13] So we can imagine what pressures, and what illegal takings of water, were put in motion by the most enterprising farming groups.

2. Gigantic Undertakings

Thus, in addition to the lagoon the Republic had to turn her attention to a front in an "inland war": the huge scenario of rivers great and small coursing through the hinterland. Throughout the modern era the rivers were increasingly the focus of technical, legislative and financial efforts by La Serenissima, culminating in one of the mightiest of public works for its daring, its duration, and the financial resources devoted to it.

Marco Cornaro, in the mid-15th century, was the first to have a broad, general view of the forces and trends that governed and threatened the life of the lagoon.[1] Securing control of the mainland allowed Venice to consider the dynamics of the lagoon within a broader territorial setting.

In his overall vision, the rivers and the gradual filling they caused played a leading role. Of course the hydrographic system had its own complex structure and made its dynamic action felt in the waters of the lagoon in various ways. In the north, for example, the Sile-Piave system influenced river travel between the city and the hinterland, and it caused the spread of reeds, a consequence of fresh waters.

Very different and much more serious where the lagoon was concerned was the role played by the Brenta system. A rushing river throughout its entire course, defined by a 17th-century hydrographer as "the chief source of our many hardships, always changing in its wide bed,"[2] it was feared not just by the citizens of Venice. For centuries it had been the object of rivalry and armed conflict with Padua, and had been used as a weapon for war and revenge between the two cities. This raging river, the wildest in Veneto, emptied into the middle lagoon causing immediate and visible damage to the salt water and the depth of the lagoon basin.[3] For at least four centuries taming that river was the main focus of theoretical and experimental efforts by the greatest hydrographers of the time, who made significant contributions to founding modern hydrography, leaving in their records a vast literature on the subject.[4] After various interventions in the 14th and 15th centuries diverted the outlet some distance away, some of them large-scale diversions—such as the one made in 1457 at Santa Maria di Lugo—there finally evolved the ambitious project of diverting the Brenta outside the lagoon, first conceived at the end of the 15th century. Between 1501 and 1507 the Dolo dam was built, whence the Bruson diversion channel took the river as far as Conche, where together with the Bacchiglione it emptied into the Montalbano canal. This was an impressive project from both the social and technical viewpoints. It required expropriating land from many owners along the course of the river, widening the bed of the Brenta by hand excavation, and finding new channels for the countless torrents deprived of their outlets because of the diversion. The solution, which certainly had the merit of saving the lagoon from filling in, proved defective in time because new problems began accumulating to the south, near the barrier island of Chioggia, which plainly showed increasing damage to its inland waters.

By then it was becoming more obvious to the Venetian specialists that a process was needed for delimiting the territory of the lagoon—it would take several centuries—whereby Venice gradually excluded with massive levees the rivers that once emptied into it, at the same time isolating the salt waters from the mainland. As the 17th century began, the "lagoon boundary," the dividing line marked with stone pillars that would ultimately extend from Chioggia to the northern end, already symbolized the Republic's new level of technical and political awareness. Mainland interests were not those of the inland waters, and the latter had to be defended at any cost, even against the needs of the *land-based* economy of the other towns. Some of these interests were considerable, but always considered strategically "inferior" and subordinate to the ultimate goal of preserving the lagoon.[5]

With the mighty project of digging the New Brenta *[Brenta Novissima]*, completed in 1610, many of the drawbacks were resolved that had threatened the Chioggia lagoon, and the uncertain frontier between inland and mainland waters was more clearly defined: the southeast bank of the new waterway was now the closed border of the lagoon, and beyond it no farming or other economic activity might compromise the free flow of the waters.[6] This solution, together with later interventions and constant secondary improvements and modifications, finally proved decisive at least for a few centuries, since it prevented the most dangerous river emptying into the lagoon from producing its harmful and, in the long run, fatal effects. Later, under Austrian domination in the mid-19th century, technical concerns assailed the Venetians as they watched the steady deterioration of the lagoon—incompetence sometimes won out over reason and thoughtful observation—and the banning of the Brenta was called into question. In 1840, after a devastating flood the previous year, the outlet of the river was returned to the lagoon at Chioggia, where in fifty years it caused filling measured at 34 square yards. In 1896, a new channel from Conche to Brondolo and through the outlet of the Bacchiglione, which took it directly to the sea, once again banished the Brenta from the lagoon. This is the solution that withstands the test of time today.[7]

It was in the 17th century that the great projects for channeling or diverting rivers were carried out on the broadest scale. Various schemes and projects, some conceived in the previous century, were put into practice. Among them was the effort to control the northern drainage system dominated by the Piave and the Sile. To limit damage from the Piave, an important river for inland navigation and the cause of frequent floods, a mighty channeling project had been carried out in 1543—the San Marco levees—along with various diversions to channel floodwaters. Only a century later was the river slated for large-scale diversion from the lagoon. In 1664 the Piave was diverted so as to discharge at the mouth of the Livenza at Santa Margherita, in an attempt to protect the lagoon from a great year-round inflow. Not even this solution was without drawbacks. On the contrary, two centuries later a great engineer would even call it "unnatural" because the flow of the river was reversed.

Levee breaks and floods were frequent and sometimes disastrous in the area, in the years immediately following. Even though a half-million ducats had been spent on the project, an expert recalled in the 1680s, a number of neighboring fields were "turned almost completely into stands of Reeds, while Trees and Vines died." Then a great flood in 1683 caused the river to empty into the port of Cortellazzo, where its outlet remained. So an old project of Sabbadino's was in effect realized, in obedience to the irresistible forces of nature, at the same time providing a lasting solution to the problem of distancing the river from the lagoon basin.[8]

Meanwhile the Sile, a quieter and clearer river, feared only because of its fresh waters, was diverted into the empty bed of the Piave, perhaps restoring the ancient hydrography that once governed the lower courses of both rivers.[9] In those decades even an old arm of the Po, the "Venetian Po," which had caused a memorable flood in the 12th century, the "Ficarolo break," began creating new and unexpected problems for the lagoon. Four centuries after the break, because the Reno emptied into it near Ferrara, the great river had always sent more and more of its waters into the Venice arm. The last stretch was called the "Furnace Po" and outside the shore barrier had formed three branches named for their respective directions: North Branch, East Branch, and South Branch.[10] The first of these branches especially had begun to produce plainly visible degradation in the lagoon. "I declare," said treasury lawyer Filippo Zorzi with alarm at the end of the 16th century, "that the great harm the river Po causes through its North Branch, and its threat to property in this city and the entire State, silting and filling the ports, cannot be described or estimated by human judgment."[11] Even allowing for rhetoric and exaggeration, the threat was not to be underestimated.

This branch was gradually silting up long stretches of coastline to such an extent that, as an engineer wrote in 1662, almost sixty years after the Po had been permanently diverted, "in 50 years it has changed about 15 miles of Sea into land, and where ships once passed, men have built their dwellings; and since shelter is denied the fish, greater space is given to animals."[12] This huge arm of the river was now pressing against the lagoon and was plainly silting up the ports, Fossone and Brondolo most of all, but Chioggia, Malamocco, and Venice as well. Prompt and resolute action was required, and the magistrates acted. The Po itself was permanently diverted with the mighty "Porto Viro cut" in 1604. It was a daring operation, but also providential and farsighted, and it allowed Venice to protect the lagoon from obvious environmental harm and degradation. Some specialists asserted a few years later, "the Po cut project has not produced the positive effect the public expected from diverting the North Branch of the Po from the Sea Inlets and the Lagoon."[13] Nonetheless, the project was completely successful. Three centuries later, observing how the coastline had evolved under the action of that branch of the Po, and how it had changed since the diversion, a specialist could declare that the present state

> plainly showed how disastrous it would have been, not only for the port and the lagoon of Chioggia and the outlets, courses and watersheds of the Adige, the Brenta and the Bacchiglione, but for the entire system of the Venetian lagoons, not to employ so providential an expedient as the Porto Viro cut.[14]

3. Sunset for the Republic

In the 17th century one river authority, the one overseeing the Adige, came to acquire more resolute and active functions: already in 1518 after many devastating floods the river had been placed under the direct supervision of three magistrates. On 7 September 1677 the Overseers for the Adige were established, charged with seeing to its navigability, inspecting and repairing the levees throughout its course, building and maintaining water-control works, and controlling the use of water for power. As customary, they were assigned broad competence in matters technical, financial, and administrative concerning the entire course of the river in Venetian territory, sometimes in concert with the Authority for Waters. For managing and controlling the use of primary resources like water in the hinterland, the Republic kept extending its regulations on various levels of enforcement: intervening technically, creating authorities, issuing and verifying standards. The same principles that had guided intervention in the lagoon were now applied to a much wider territory, in an attempt to protect and reinforce community rights and equality before the law. In fact these rules represented the very conditions for developing community life based on the enjoyment of natural resources. They opposed precisely the kind of events so effectively described in a law of 26 October 1658: "With extreme injury to public interests, as well as obvious damage to individuals, many persons exercise control over rivers, levees, and waters without authority, placing such obstacles that travel and navigation, the uses and flow of the waters are blocked."[1]

Once again water use was recalling the obligations of community living. Violations by individuals were not acceptable because they would entail the loss of enjoyment of essential rights for the great majority of the population. At the same time failure to obey would open the way to serious damage and compromise the entire system, canceling the rules of a collective game. Without limits to the private appropriation of resources the private citizen's interest tended to destroy itself. For the rivers as well, the great political lesson applied in governing the lagoon—considered symbolically and in fact as the ultimate reality to which everything must be subordinate—came to be stated as a principle of "equality in prohibition," the other side of the coin being "regulated freedom" for all. Once again, in affirming an absolute limit the Venetian ruling classes set themselves as the unquestioned authority, a universal regulatory state whose existence and vigilance were essential for the unfolding of community life itself.

When the Republic fell, in 1797, her grandiose governmental structure of standards and norms collapsed as well, a system built over centuries of political autonomy for dealing with her countless partners in a vast international market and with a habitat that was one of a kind. Some large projects were carried out around the beginning of the 20th century, especially for improving the sea inlets at Malamocco, Lido, and then Chioggia. Under Austrian government, Venice had been privileged because she was a port on the Adriatic. But in those days sails were being supplanted by great iron ships. Nonetheless, Venice's shipping was "gradually being transformed from 'seagoing' to 'local and coastal'."[2]

But the battle against the natural forces and undisciplined social pressures that threatened the life of the lagoon had been won. Unlike the northern Adriatic lagoons, which disappeared as a result of uncontrolled forces, Venice had become mistress of her fate.[3] Nature and its economic uses had been successfully bent to the ultimate goal of Venice's survival. Of course the victory was no more lasting than any contest against natural and human adversaries can be. Moreover, politically fallen and defeated by history, Venice was forced to hand over her destiny to others.

Of course the city tried as best she could, under the various governments she was destined to have, to preserve the tradition of laws and prohibitions developed over centuries for safeguarding the lagoon. They are represented to some degree in the Austrian regulations of 1841, the Italian laws of 1936 and 1963, and the special legislation of 1973.[4] At least the city has maintained and imposed on her various governments, with varying degrees of success, the inescapable fact of her fragile difference. But the political soul, the "religion of good government" that inspired the Venetians' legislation and political conduct, was fated to disappear.

The urban community, the classes and groups that gave it life with their countless associations, marking places and seasons with festivals, celebrations and rituals, have disappeared forever. Sometimes cities die in the depth of their souls, even though outwardly everything stays unchanged. Of course that superior civic virtue has been lost, the spirit that even in critical moments when the regime was changing led the city authorities to concern themselves even with "irresponsible neglect of dogs in the city," ordering "all Citizens keeping Shops" to leave "accessible all day on the Public Way a Pail of fresh, clean Water."[5]

V

Decline

1. The Revenge of the Land

Throughout the contemporary era, and most rapidly during the last fifty years, Venice has undergone profound changes that have impacted the land and the entire lagoon habitat, thoroughly modernizing its economies and changing the profile of the social structure. Once fallen from her preëminent, autonomous position as a city-state, Venice saw growing around her the economic pressures and the territorial interests of what had been her hinterland. She could only look on helplessly as a centuries-old course was reversed. Whereas once the city-state subordinated to her interests the waters and lands that surrounded her, now it was her turn to undergo the pull of outside forces tending to engulf her in a land-based economy. It's also true that the Adriatic, as we shall see later, was having its revenge. At that time, Venice was opening more to the sea by widening and deepening the inlets, facilitating access to the lagoon for the great iron ships. The new vehicles of maritime trade demanded ever deeper waters.

For that purpose, after the Republic fell construction was undertaken of the seaward jetties intended to favor the deepening action of the waves. Better scientific knowledge of the behavior of sea water, starting in 1806 (still under Napoleon), suggested more effective solutions for keeping the inlets deep and safe. The gradual obstruction of the inlet at Lido caused efforts to be concentrated on Malamocco, where long parallel jetties were built—chunks of Istria stone piled up loose—forming a sort of channel in the sea. Inside it, the wave motion would constantly erode the bottom and the sandbanks that had become established over time. The success of this operation, planned by Pietro Paleocapa and begun in the 1840s, heralded the improvement of the remaining inlets to the lagoon. In the course of the century, at the initiative of a united Italy, the "channels" of Treporti, Sant'Erasmo and San Nicolò were combined, creating the single inlet of Lido, and in the next century a similar technique was used to improve the inlet at Chioggia.[1]

Between these opposing but essentially converging pressures, Venice was increasingly influenced by the land. New technical aspects of world trade and new "outside" political pressures tended to subject the city to processes that were foreign to the balance of a habitat that had been laboriously built and was easily damaged.

The loss of political freedom—the very freedom that the official rhetoric of the authorities stipulated as their ultimate purpose in protecting the lagoon—had immediate repercussions on the condition of the city, and on policy concerning the lagoon. Under the French, serious harm was done to the artistic heritage, especially in the islands, with the suppression of churches and convents, the transformation of squares large and small, and the demolition of buildings that needed restoration. French domination of the city began in 1797 and was characterized by the plundering of art works and interference with the urban structure that would leave its mark on the entire 19th century. Again, with the suppression of the old authorities came heavy fiscal pressures from the invaders, the decline of maritime trade, the move to the mainland of some of the city's old patricians, and a steady, rapid process of decay on the island. According to a respected historian of the city, it was "during Napoleon's domination that Venice came to resemble a city in ruins, which long fascinated the romantics."[2]

The abandonment and decay that hit the city in the 19th century resulted as usual in damage to the lagoon, with expropriation of public spaces by private individuals "very frequent in those years."[3] This was done mostly by filling in *rii*, the countless little canals scattered everywhere in the lagoon. It was not just the spontaneous, pervasive result of illicit actions in closing areas and channels to traffic.[4] Such a tendency, difficult to quantify, is understandable once the old legislative apparatus had become an empty shell, and the power and prestige of the Republic's authorities were only a memory.[5] But the reality did not stop there. In those years the tendency to take over space on the shores of the lagoon was seen as a new form of enterprise, and it chiefly involved significant social groups in the city. In that phase, as has been recently pointed out, those responsible were "various owners of real estate, who sat on the City Council or held other positions of influence, strenuous advocates of filling canals because they would avoid the taxes (the notorious *grossi*) that had been paid since the earliest days of the Republic for dredging and maintaining the canals in front of their houses."[6]

The feeling for the fate of the lagoon in many parts of Venetian society had changed. The new Austrian domination, although less openly bent on plundering the city than the French, did nothing to remedy tendencies in progress. Of course, some "lagoon policy" initiatives were necessary in those decades, beyond keeping the lagoon from becoming the scene of an unbridled private takeover. In 1841 the Austrian government issued "Regulations preventing the damage being done to the Venetian Lagoon," legislation that for almost a century afforded protection, and in any case reasserted the unique nature of the inland sea.[7]

Nevertheless, in those decades the city had fallen into a deep economic depression, especially in the twenties. Hit by famines and epidemics, as in times of crisis long ago, she saw her population shrink and become impoverished.[8] The weakening of the city's ruling class, like the influence of mainland interests, had a great impact on policy decisions concerning the lagoon. Precisely in those dark years in the economic and social life of the city, the lagoon was perhaps at the greatest risk in all its history. In Vienna, of course following pressures and protests from some quarters in Venice and on the mainland, the Austrian government was about to consider restoring the rivers to the lagoon basin, a decision that would have abolished centuries of mighty water projects carried out by the Republic. The damage caused by diverting a few rivers in the mainland countryside had convinced some specialists that the diversions carried out by the old authorities had been a costly and pernicious mistake. Moreover, no intervention had ever been made in the previous decades and centuries without provoking criticism and opposition. But now for the first time it was a foreign power marshaling the opposition on behalf of a decision of incalculable consequence. In the early 1820s, as Pietro Paleocapa recalls in his brief account of the debates and events, while awaiting the final technical decision from the government in Vienna it was "considered certain that it would not fail to arrive, and so there spread in Venice what may well be called a sense of terror."[9]

But in that century, aside from technical errors in judgment and the scattered, unruly interests that were regaining the upper hand, the "land forces" had acquired irresistible power, describing themselves as the bringers of "progress." It was hard to restrain them. In 1846, for example, midway through the Austrian domination, a railroad bridge crossed the lagoon and linked Venice to the mainland for the first time. This event had great symbolic significance, and it caused profound changes in the geography of the lagoon and its relationship with the city. Venice was torn from her centuries-old isolation, her proud maritime setting, and forced to share the destiny of ordinary towns. It must be remembered that at the time "normalizing" Venice was part of a positive trend towards development in the entire area. Building the bridge brought with it a vast draining of shores and channels that eventually increased the land area at the expense of the waters. It also represented one aspect of a more general process binding Venice to Italian territory. Completion of the rail line between Venice and Milan in 1857, along with the building of other links with the provinces of Veneto, made Venice more and more of a typically land-based attraction.[10] This required new projects to meet the growing needs for space and structures, created by economic and commercial polarization. So the city struggled to regain her economic dynamism, which struggle however tended to relegate to the background the ancient rules of her inland sea.

We must also remember that in the 19th century the forces of the mainland economy—never dormant even during the greatest periods of the Republic—put forth arguments and interests that were often legitimate, but more openly in conflict with preserving balance in the lagoon. In that phase, if time vindicated the specialists who had determined to banish the rivers (*proving* in fact that the lagoon had been saved), the same was not true for landowners and dwellers in the hinterlands. The same physical processes linked to the stream diversions were now showing the negative signs we have mentioned in the preceding pages. As decades became centuries that solution, which gave the rivers a longer and slower course parallel to the sea, had raised their beds above the level of the surrounding countryside. So the farmlands were less and less able to drain off rainwater and tended to become marshes, reduced "to the wretched state of fishponds," in the terms of an expression then current.[11] The marshy, malarial landscape typical of land below sea level came to characterize what observers called the vast "Veneto estuary," the whole broad stretch of shoreline along which the countless Po tributaries emptied. At the edges of the lagoon, closed to fresh water by the boundary levee, land reclamation was already increasing at the end of the 19th century, with numerous landowners using pumps; they often wound up draining land closer and closer to the lagoon, when they didn't manage to spread the hydrographic disorder of their lands into the waters of the lagoon itself.[12] This pressure as well, despite the sometimes understandable interests that caused it, revealed the increasingly land-based logic of the economies around the lagoon and the city. The defenders of city and lagoon sometimes included representatives of the so-called specialist class who for professional reasons, because of their cultural background and historical memory, still had a special and unquestioned cult for the inland waters. A ringing testimony is this oration by a valiant engineer at the end of the century:

> But at this point someone may ask, Why such zeal for conserving the lagoon? Wouldn't it be better to hasten the filling and turn it into farmland? Under the Most Serene Republic no one would have dared to ask that question for fear of being judged an enemy of the State and a traitor, and surely no Venetian would have the courage to propose it today, at least after study and reflection; but since few these days are concerned about the most unusual condition of our city and the other sizable settlements in the lagoon area, and few are informed of the amazing efforts whereby our ancestors succeeded in guaranteeing for both a lasting existence, in this case I cannot abstain from replying.[13]

And our engineer replied in kind, showing by a historical survey of the Republic's water policies, as many of his predecessors had done, what the fate of Venice would have been without the work of preserving the lagoon.

Meanwhile, the push to reclaim the new marshes and malarial habitats still found an active advocate in the Italian State. From 1900 on, reclamation laws granted future advantages to landowners organized in consortia operating at the edges of the lagoon area.[14] We must say that in the last analysis the lagoon managed to fend off the "reclamation fever" that raged during the whole first half of the century and could have inflicted irreparable damage on it. Even though, as we shall see, in this phase the lagoon suffered considerable reduction in land and water area for other purposes, Venice's prestige, her legislative tradition, and the historical memory of at least part of her ruling class, were certainly an obstacle to bold plans for tampering with the lagoon for agricultural purposes. That was no mean achievement. The philosophy of "territorial conquest" characteristic of the time created a climate quite inclined to make exceptions to the law and scorn the past. And in some cases the "state of emergency" created by uncontrollable events inspired choices that became examples and precedents that were surely pernicious for the future of the lagoon. Reclamation was finally done wherever it proved difficult to restore to their original state changes made within the "boundary" of the lagoon. This was true of the Chioggia lagoon area, which had silted up extensively after the Brenta was allowed to empty into it from 1840 to 1869. "Given the irreversible nature of the phenomenon," wrote Silvano Avanzi, "there was no longer any reason to maintain the lagoon restrictions on the new lands inside the boundary drawn in 1791. On the contrary, since they were marshy and malarial, they had to be exempted from the restrictions urgently, in order to reclaim them." Therefore, disregarding the Austrian Regulation of 1841, two decrees (in 1924 and 1926) redefined the area and required the landowners to form a Consortium to carry out the work.[15]

2. Rural Wealth and Urban Poverty

After Veneto was annexed to the Kingdom of Italy, in 1869 construction of a new port was begun at the western end of the Giudecca near the end of the railroad bridge, the Maritime Terminal that would mark a new commercial renaissance for the city.[1] Of course, among the processes impacting Venice and her territory now they were part of the new Kingdom, we must discern trends and outcomes that were not always simple. Indeed, the rise of a real industrial center, encouraged by the coming of the railroad and the Maritime Terminal, bound Venice to the mainland, but it also revived her commercial and seagoing calling. It was the strong land-based economy that finally created the need for a larger and better-equipped port. So in 1917, in the midst of the war, a decree from the military governor gave the force of law to an accord between the city of Venice, the Italian State, and the Industrial Port Corporation [*Società per il porto industriale*]. That was the origin of the new port of Marghera. It was a victory for a group of manufacturers and financiers, and it was to be the focus of many polemics over the cynical exploitation of public resources for private gain, which even adopted the rhetoric of a seagoing renaissance for La Serenissima. For its promoters, this new port would allow "Venice to 'recover' her dominance in the Adriatic and help raise Italy to the rank of leading Mediterranean power."[2] However, linked to that initiative was the creation of a great industrial center, meant to restore economic importance to the city in the course of the 20th century. But maritime expansion and industrial concentration without a policy for the lagoon turned to the detriment of the inland waters, further subjecting the amphibious city to the requirements of life on land, and to domination by capital and profit, a concept unknown to the magistrates of the Republic. As we shall see, from the industries that clustered around the new port of Marghera would come the greatest damage and the deadliest menaces to the health of the lagoon habitat and the city.

Of course, the historian may not judge the actors he is studying solely in the light of their awareness of a problem that has again grown acute in the culture and the needs of his own time. They were obeying necessities different from the ones that inspired the men of the Republic. But we must add that after the long and humiliating neglect Venice suffered at the hands of the French and the Austrians, many in the ruling class saw another specter as frightening as the degradation of the lagoon: the city's economic decay. As a contemporary still kept repeating in the early 1920s to anyone who feared the degradation of the city, "We cannot and we must not let our Venice be perpetually doomed to die." And he recalled, not without reason and good sense, what a different situation his contemporaries must take into account:

> In what I shall call the traditional and static context of the old world, complete closure was for Venice a protection as well as part of the rather mysterious fascination that surrounded her. But today? In radically different circumstances, in a world teeming with communication and progress, isolation would smother her, and her name would be added to those of other famous dead cities.[3]

The problem of making the city live from new economies was certainly real. But surely the historian can hardly be charged with anachronism if he expects from his 20th-century heroes a more enlightened sensitivity in reconciling the needs of the economy with those of the lagoon habitat. As we have already seen, generations of their ancestors had undertaken the task, with no small measure of success.

So the 20th century would simply continue along a route already marked out. In 1932, in the midst of the Fascist era, Venice's isolation was again violated with the construction of a highway bridge two and one-half miles long, the Lictor Bridge, since renamed the Liberty Bridge; it further strengthened the city's inclusion among places on the mainland.[4] This was a transformation of great cultural significance, especially since it made Venice accessible to the new vehicle that to a certain extent defined the new individuality of 20th-century transportation, the automobile.

Along with these urban modifications, changes were made to the lagoon without any goal of conservation. They were dictated by outside economic interests, and they were often harmful to the lagoon's fragile balance. While in the 30s new channels were dug (Victor Emanuel III and Malamocco-Marghera, the "tanker channel") to answer eminently land-based industrial interests, within the lagoon the water area was shrinking. Starting with the new century reclamation work proceeded relentlessly, leading to the gradual filling of mud flats, salt flats, and channels. This was done almost everywhere: near the delta of the Brenta, at Lugo, Montiron, Ca' Deriva, Veronese de' Marzi, etc. In the first sixty years of the 20th century nearly 9,880 acres were drained. But the city, with her growing functions and economic ambitions, required still more land area. Thus, 1,730 acres were filled for building the airport of Tronchetto, and extensive filling was done between 1917 and 1963 to create space for the three industrial zones. Over 7,400 acres of the lagoon basin were put to other uses.

> Overall, it has been observed, the work done in the lagoon from 1900 to 1973 cut off or removed from tidewater a good third of lagoon itself, while dredging removed over 69,000,000 cubic yards of the bottom.[5]

Thus the land-based economy tends to take away space from the water. Situated as she is on the territory of a nation-state, Venice can no longer think of herself and her economies as being inseparable from her lagoon. And the lagoon has lost its inviolable sovereignty.

No less significant has been the city's population curve. Despite a significant increase at the beginning of the 20th century after its 19th-century decline, Venice has undergone a substantial reduction since World War II. From 1951 to 1981 the urban center went from 175,000 to 92,000 inhabitants. This was a genuine exodus of Venetians, some of them long-time residents, limited in its effects only by a rapid process of demographic substitution. New families filled some of the vacancies left by the many who abandoned the lagoon. That decrease hid the loss of about 150,000 individuals in thirty years, over 5/6 of the population, replaced in part by immigrants who modified its social composition. The population was new, and older: in those years the average age reached a peak of 43.[6]

This is surely one of the most significant signs of the economic and functional changes that Venice has undergone in recent decades. They are consistent with the changes seen in other Italian and European cities in the latter half of the century, although they have not been so destructive or disruptive of the urban social fabric here as in other European centers.[7] But they have certainly had peculiar effects on this city, situated as it is in such a special and fragile habitat. The ancient trades and crafts like farming, fishing and retailing have declined in favor of the service and public-sector economies. The increasing specialization of the city as a leading tourist attraction, a perennial goal of national and international pilgrimages, has inevitably introduced a profound change in the ancient social balance. Retail prices and rents have finally driven workers and laborers from the city; they cannot afford the exceptionally high cost of living. This in addition to a long-standing shortage of housing on Venice's island.[8] On the other hand, every city with a strong identity is burdened with a monolithic image. Venice is doomed to be admired by foreign eyes for the splendor of its matchless structures, but no one sees the wretched housing of its poor sections or the dilapidation of its lower-class districts. And yet the decay of "ordinary buildings" was one of the wounds that had the most serious consequences for the health of Venice. According to a Minister for Public Works at the end of the '60s, this was "the chief cause of the demographic decline of the city."[9]

The *popolo minuto* has all but disappeared: the shoemakers, tailors, tinsmiths, woodworkers, barbers, masons, etc., the crowd of social types that truly give life to a city and ensure its everyday routines, its constant "social maintenance." The city tends to lose the services that make possible the very survival of its inhabitants.

The population of Venice, permanent or temporary, has become chiefly one of executives, entrepreneurs, professional people, retailers, students, seasonal residents, etc.,[10] who have a somewhat limited and partial relationship with the city, at least with that particular dimension of the city that is not defined by the buildings alone but includes the lagoon as well, as part of an overall habitat. The new citizens have no relation to the water except as a means of transit, a spectacular part of the urban landscape, or an obstacle to the frenetic movements that are part of daily life today. Demographically and socially the ancient island has slowly become a kind of artificial world in which there is hardly any connection between the citizens and their habitat. The old urban community has dissolved. While many of Venice's ancient houses stand empty and some of them are abandoned, young people and workers live in towns on the mainland, in Mestre or Marghera, come to the city by the thousands each morning for work or study, and then return home.[11]

The boatmen, fishermen, freight carriers, and dredgers have largely disappeared, the crowd of ordinary working people who constantly saw to observing and maintaining the city and its setting in the lagoon.

The ancient symbiosis between the city and her people on land and water, the foundation for the conservation miracle to which Venice owes her survival, has dissolved and been replaced by a set of relationships that are occasional, fleeting and superficial. In its place—also because of the profound social modifications that have impacted society in our time—has appeared the anonymous "urban crowd" that uses the city as a backdrop, sumptuous but alien to their material interests and their impatience, and experienced essentially with indifference. Interest in Venice is only esthetic: thousands of tourists come and admire her for a few days or a few hours, then go away. Even the cultural bond between the Venetians and the water is profoundly changed. Something has broken in the soul of the ancient community, something that escapes superficial observation. The countless canals that intersect with streets and alleys, buildings and squares, forming the old regime's ancient and peculiar thoroughfares, are now seen as a handicap, an obstacle to the hunger for speed that consumes city dwellers in today's world. Venice is not exempt from global trends. Moving around the space of the lagoon according to a time scale dictated by our mental enslavement to the insane commands of faster and faster producing and consuming relegates a slow, human relationship with the water to mere inefficiency, a "waste of time."

This transformation of classes, people and cultures is a prime cause of the decay that has impacted the lagoon and is undermining it today. In an utterly peculiar way with respect to the past, Venice is still the exception today. Governing her cannot be entrusted to the prevailing rules: like the natural forces in the lagoon, they tend to destroy her. As in the past, an extraordinary degree of commitment and creativity is required of the political powers to overcome them. As always, governing Venice is a challenge that sweeps aside everything "normal." It represents a great laboratory within the western world: it calls for inventing a government that can restore life to the city and rebuild its living community. Reaching this goal requires disciplining the lawless, contradictory forces that are making Venice into an inert museum city, an increasingly debased receptacle for industrial pollution.

3. Political Decay and Industrial Giants

Not only the waters have been subjected to such extensive manipulation. The mainland has seen changes too, indeed gigantic transformations. As we have noted, between the World Wars Marghera saw the construction of a huge port and then a first-rank industrial complex that profoundly transformed the local economies. We may say that owing chiefly to the industries located in the area Venice has rediscovered in a new form her onetime economic primacy in the Adriatic, enjoying an unprecedented prosperity after her long, relentless decline during the contemporary era. A new dynamism in enterprise and social classes hit her from the outside at close range, so to speak, bringing work and incomes to areas that once were under her absolute control.

Industrial expansion at Marghera, however, increasing in the postwar period, has not been without consequences for environmental balance in Venice and is undoubtedly one of the dilemmas facing the city today.

Most of all the predominance of the chemical industry, with Petrolchimico's plants and related activities, constitutes a profoundly alien presence, in conflict with the nature of the location, the new roles of the city, and the needs of the environment in which Venice stands and must survive. We must remember that despite the destruction and serious degradation of the last decades the lagoon still constitutes a singularly original and rich habitat for flora as well as fauna. Besides many common species of trees, there are numerous varieties typical of wetlands, such as eelgrass, spartina grass and limonium. Fish species have surely diminished since the old days, but they are still extraordinarily abundant. In the woods there are otters, polecats, weasels, and wild hares. Birds are the chief natural wealth of the lagoon habitat. From the red heron to the royal swan, from the wild duck to the spoonbill, from the mallard to the widgeon, a great number of fowl, both migratory and nesting, inhabit the lands and the waters. It has been calculated that between 1976 and 1992 an average of at least 40,000 birds wintered in this area. This is an uncommon bird population, making the Venetian Lagoon one of Europe's most important wetlands for waterfowl. Not by chance is it listed as one of the four most important zones between the Mediterranean and the Black Sea: the Guadalquivir delta (Coto de Doñana) in Spain, the mouth of the Rhône (Camargue) in France, and the Danube delta in Romania.[1]

Such natural wealth, together with the priceless artistic heritage contained in Venice's buildings, makes the presence of the Porto

Marghera industrial complex on the mainland shore increasingly foreign and discordant. We shall not indulge here in facile esthetic considerations. No one can deny that the plants in the area have produced wealth and given employment to a large numbers of families for several decades, even if often accompanied by serious alterations in the health of the workers or the natural balance of the surroundings. And yet, for several decades the complex has appeared from our standpoint to be a development option that is profoundly alien to the environmental and historical realities of the city in the lagoon, part of an industrial culture that is singularly insensitive to problems of natural balance as well as the quality of urban living.[2] Fragile habitats and cities that are works of art can no longer coexist with polluting industries.

Petrolchimico in Marghera is the center of economies that profoundly alter the quality of the waters in the lagoon. It's the heart of a complex numbering about 150 operations in petrochemicals and light industry, shipbuilding, and oil refining: 50 of them produce liquid wastes in significant quantities and discharge them into the lagoon.[3] It is calculated that nearly 5,000 ships enter the port each year, almost 1,100 of them carrying petroleum products. Bear in mind that until the mid-'60s the oil tankers went through the San Marco channel from Lido; in other words they entered the city, bringing their cargo of crude into one of the most fragile and delicate urban jewels on earth.[4] Now, in addition to the tankers, barges transfer oils and other hydrocarbon derivatives to the large ships anchored in the port.[5]

So it's inevitable that residues, waste, and discharges from highly toxic industrial activities should wind up in the lagoon. Ongoing analysis of the waters has recently shown the presence not only of hydrocarbons and their derivatives, but also of lead, copper, zinc, mercury, and according to numerous reports, even dioxin.[6]

As environmental associations emphasized at the beginning of this decade, presenting official documentation of the problem: "Venice, in particular the area of Porto Marghera, is the largest center in Italy for the disposal of toxic wastes, especially for incineration." The lagoon has been turned into a private dump for industry.[7]

We must remember, in connection with these statements, that in recent years a new enemy of the lagoon has joined the old ones. As the reader has seen, in his time Cristofaro Sabbadino identified the most dangerous enemies as the sea, the rivers, and men. Today he would have to list a new, insidious enemy of the inland waters: the sky. Smoke and gases produced by the industries of Marghera, together with the exhaust from automobile traffic, fall into the lagoon with the rains, bringing another load of toxic materials to be added to the poisons from the land.[8]

On the other hand, industrial activity is not the only source of pollution. Remember that beyond the lagoon lies a "catchment" of about 725 square miles of land that sends rain and river water into the lagoon basin. About 31,783 trillion cubic feet of fresh water flow into it each year from the rivers that still empty into it. The rivers, rainwater, and extensive agricultural activity—especially corporate corn growing—contribute significantly to degrading the lagoon, as they are increasingly loaded with chemical residues. Now that it's become an industry, even the ancient productive activity of the countryside, even stock raising, makes its own special contribution of poisons.[9]

The consequences of these polluting pressures are easy to imagine,[10] especially since they come on top of city discharges, which have at least tripled since 1900. Within the "catchment" there are 101 towns, which in the 1991 census numbered 1,461,000 inhabitants. Therefore, the waste discharge is not only specifically industrial but to a great extent human, and it is only partially filtered by the existing treatment plants.[11]

The inevitable result of productive activities that have become ever more alien and hostile to the waters of the lagoon, insofar as they degrade the natural habitat, is generally recognized as being serious. Proliferation of algae, fish die-offs for lack of oxygen in the water, and a profound degrading of plant and animal life in the lagoon are among the best-known effects.[12] Never have economic activity and the health of the environment come into such violent conflict as in this phase of Venice's history.

But this aspect of industrial and human pressure from the mainland is not the only menace that is causing great alarm. A further serious threat hangs over the lagoon: the risk of a serious accident that could dump colossal volumes of oil into the lagoon, with unthinkable consequences for the city and its structures, and the whole lagoon habitat. That risk is not really remote, since in early December 1995 a break in the underwater oil pipeline that links Marghera to the barge terminal nearly caused an environmental disaster.[13]

So the constant pollution, damaging and sometimes destroying trees, killing or driving off fish and birds, also contributes to making the action of the waters more polluting and corrosive to the foundations of houses, the underpinnings of buildings, and the outer surfaces of all the city's structures. And yet this all-pervasive degradation is only the proven aspect of the damage that is being inflicted slowly and will bring about collapse in a period of time that remains indefinite. Like a huge shadow the daily poison, the enemy working day after day, brings a deadly menace to the city and her waters, a threat that chance or human error could turn into a catastrophic reality.[14]

VI

Saving Venice

1. As in the Beginning, Threats From the Sea

Today as in the past, the greatest dangers come not only from men but in some respects above all from the sea. With the great rivers banned that once silted up the lagoon basin, the most serious threats to Venice's preservation and her very survival now come from the Adriatic. The increasing risks and injuries to the city caused by wave action appear in the worsening phenomenon of "high water," a rise in tide levels that sends water into the basilica of St. Mark's, invading streets and squares. St. Mark's Square, which at the beginning of the century was flooded five to seven times a year, is now under water 40 to 60 times a year. The Adriatic tide, the greatest in the Mediterranean basin, has changed radically compared to the past, not only causing ever greater periodic distress for residents of the city and damage to property and merchandise, but gnawing away like a file at the foundations, surfaces and structures of buildings, the bases of monuments, and the ground under squares and streets.[1] Already ten years ago it was observed that the mean high tides were rising above the Istria stone, the impervious foundation of Venice's buildings built on wooden pilings, and attacking the porous, absorbent upper courses:

> The intertidal zone is now directly in contact with the brickwork. Brick, and to a greater extent tile, is extremely porous, and the saltwater attack is taking on worrisome dimensions. As the salt crystallizes it expands, bricks scale off and tiles crumble, the mechanical bond weakens, and structural fragility increases.[2]

Moreover, besides the increasing frequency of high water, the city is more frequently threatened by severe flooding. One event is a milestone in the recent history of Venice. On 4 November 1966 an unusually violent storm devastated the barrier of breakwaters in several places and flooded the whole city. The flooding was unexpected and disastrous, perhaps one of the most "catastrophic" in Venice's history. In words echoing the extraordinary emotional impact of the event, one historian wrote, "It very nearly destroyed a thousand years of Venetian civilization in a single blow."[3] The water rose 6.36 feet above sea level, devastating all the stores, invading ground-floor dwellings, and damaging workshops, libraries and warehouses.[4]

That was an absolutely exceptional event, of course, but it signaled the extent to which the degradation of balances in the lagoon had progressed, and the urgency of taking action. How could two days of rain and sirocco raise the waters to a level never experienced in the past? It was immediately realized that some of the ancient seaward defenses had also given way. The glorious *murazzi* built along the Adriatic had been breached in several places and overwhelmed by the waves. One cause was the force of nature, of course, but another was man's shortsightedness.

As was reported in the aftermath of the flood, the murazzi had not been periodically reinforced and rebuilt, as they always had been in the past, since the 1930s at least. Practically no maintenance or repair had been done. In the Pellestrina area, for example, it had been decades since the "rock men" *[sassanti]* had been seen, those most typical personages in the supervision of that stretch of the lagoon:

> The sassanti were a class of specialist laborers, usually divided into two teams, who had only one task, manual but valuable. It was to bring rocks for restoring the murazzi. They were laborers, but highly regarded by the people of Pellestrina, just as we regard those responsible for our safety. The rocks used by the sassanti came from the quarries of Istria: several sailing ships each month approached the shores and unloaded the precious material. By the end of a year, about 15,000 tons of marble had been brought to "feed" the coastal defenses . . . The sassanti were careful to "drown" rocks at the foot of the reef, replacing the blocks that had been buried, eroded or carried away by the waters.[5]

The slow erosion of the defensive barriers had made the city more vulnerable to the periodic assaults of the sea. Where the murazzi had given way, new openings let in the waters of the Adriatic during storms.

In fact the processes of transformation that had meanwhile impacted the lagoon basin and its connections with the sea were more widespread and complex. Today, for example, we can plainly see the unforeseen and unintended effects of projects carried out by La Serenissima. The murazzi themselves, which were an effective and long-term solution to the problem of protecting the habitat from the sea, had another result: the sea floor in front of them was eroded by wave action, and the water grew deeper.[6] This gave more force to the waters meeting the shore.

In fact, as many studies and observations have now shown, a process of deepening erosion exists not only along the barrier islands, but in the lagoon itself. Here again we encounter, long afterwards, the more or less unforeseen effects of safeguards carried out by the Republic. The gigantic task of diverting the river outlets in modern times managed to halt the silting of the inland waters, but it had as its inevitable corollary the loss of replacement material for the lagoon basin and the seashore. That whole long stretch of the Adriatic is in fact deprived of the filling action once provided by the various watercourses, while the few rivers and torrents that still empty along the coast leave less and less material.

Even the positive action of specialists and local authorities at the turn of the 20th century potentially fostered the deepening of the lagoon:

> For decades (writes Cucchini, the chief engineer with Civil Engineering in the early 1900s) the constantly growing industries and commercial traffic have required and still require enlarging of the secondary channels, to improve passage between the city of Venice and Chioggia, the industrial centers scattered throughout the islands, and the chief centers in the bordering provinces.[7]

Enlarging and maintaining these channels was faster and easier than in the past. Instead of men with shovels there were ever more powerful dredges that removed the mud and were not careful to dump it where it would serve constructive purposes. Venice no longer thought of her habitat as a reservoir of endlessly renewable resources. Following a logic peculiar to capitalist societies, the material dredged up was dumped back into the sea, as far from the coast as possible. For decades after 1820, according to Cucchini, the material dredged up covered about 988 acres of salt flats and marshes. This practice fostered the expansion of land area that guided the Austrian government's action. But starting in 1869 the practice of dumping at sea became more prevalent, with further expansion in the first decade of the new century. In the space of ninety years, Cucchini estimates, the volume of material excavated from the lagoon was about 1.8 times that removed in all the previous centuries.[8] A huge mass of silt spread across the bottom of the Adriatic, ever more material that was permanently removed from the lagoon every year.

Moreover, the erosion process was fostered over time by the greater "clarity" of the sea water. Even the old contributions of the sea were growing progressively smaller. As briefly mentioned earlier, the construction of various outward breakwaters starting in the last century, and intended to ease access for large ships, gradually deprived the lagoon of the sediments and sands constantly brought by the sea. And even the relatively recent deepening of the lagoon for the so-called Oil Channel linking the Malamocco inlet with the industrial port of Marghera, as we have seen, has fostered a greater flow of sea water into the inland waters.

So the lagoon is gradually losing sediments, at a rate calculated to be about 1,308,000 cubic yards each year, representing an increased depth of about six inches over fifty years.[9] The process appears to be furthered by the continual passing of ships and boats along the various channels, stirring up and dispersing bottom sediments, while pollution of the water destroys the plant species whose roots once stabilized the bottom. The gradual disappearance of the salt flats, which in 1901 covered 34.8 square miles but today only 18.2, is one of the most alarming signs of the erosion impacting the lagoon today, and at the same time a cause of its steady progression.

Without mud flats and salt flats breaking the surface of the water and often covered with grass and bushes, and without the dense network of channels that lace among them (the *ghebbi*), wave action finds no obstacles in the inland waters, its force increases, and so in a vicious circle it speeds erosion.[10] The lagoon basin is tending therefore to become "flat," deprived of its inner structure, its complexity, its meanders, and its ancient natural complexity. It needs, as noted in an endorsement to the special legislation in 1973, "a process for regenerating the forms peculiar to the lagoon."[11] But without corrective intervention, without courageous initiatives that are able to restrain and discipline private economic interests in the governing style and philosophy that characterized the Republic, and without constant daily maintenance, the forces at work today tend to transform the lagoon into an arm of the sea.[12]

In this way Venice, within a lagoon that is becoming less of a barrier against the sea, is more and more exposed to the waves of her ancient friend and enemy, the Adriatic. And the risk of disastrous flooding in the city is growing steadily higher.

But a greater, more serious risk hangs over the city in our time: slow subsidence is causing her territory to sink into the water at a rate that seems to have increased in the last decades. It is calculated that in the space of seventy years, from 1900 to 1970, the lagoon area sank fully nine inches relative to sea level. This threatening tendency is now established beyond doubt.[13] But is the city sinking, or is the sea rising? Or are both phenomena combining to create an anxious situation, casting a shadow of uncertainty over the future of the city? The dilemmas that tormented and divided generations of the Republic's specialists and government officials are present again, in new forms.

It's probably the Adriatic that's rising, in connection with the overall oscillating rise in sea levels (eustatic movement), perhaps caused today by the rise in temperature of the earth's surface and the melting of arctic ice. On the other hand, the more than 6,000 wells enumerated in and around the lagoon at the end of the 1960s, together with other demands on the water table for industrial purposes, have contributed to the subsidence of the land on which Venice stands.[14] According to some specialists, this would account for the increased frequency of abnormally high tides: 24 exceeding three feet between 1920 and 1929, and 204 between 1960 and 1969.[15]

But the slow sinking of Venice could also be connected, according to some hypotheses, with a vast subsidence produced by compaction of the ground, affecting the whole broad area of the Po gulf.[16]

2. Strategic Responses and Global Threats

What has the city of Venice done in the recent past, and how have its ruling class and the national government gone about confronting the natural and man-made causes of degradation, providing defenses against the new threats? What has been planned and carried out for "saving Venice," to use the now current phrase that covers all the problems of the ancient City in the Lagoon? Overall, the answer is, necessarily, disappointing. Throughout the contemporary era very little has been done in terms of conservation and defense. Inertia has been the dominant feature, a burden on men and affairs. Perhaps we should even say that no one even *thought* of the issue, to the extent the term has any strategic content. For the succession of national governments, Venice and her waters simply did not constitute a problem. Besides, as we have briefly noted, the interventions of all kinds carried out in the last century—economic, social and territorial, public and private—certainly did not benefit the lagoon habitat. They were aimed at economic growth and its necessary territorial supports, and nothing more. In the eyes of investors and contractors, the territory was in no way special; its only drawback was that it was crisscrossed by water. The material interests that guided them often had little relation to the needs of the lagoon, when they didn't radically conflict with its ancient and delicate environmental balance. Even the customary work of maintaining the city's canals was gradually postponed and finally forgotten. As we shall see, the Cacciari administration, in the early '90s, resumed that centuries-old practice, which the Italian Republic had completely eliminated.

Only in recent years, as we shall try to show briefly, have conditions been defined—projects, standards, financing—for broad and effective intervention in the structures entrusted with the survival of the city.

Indeed, the loss of her supremacy and political greatness in the contemporary era has entailed for Venice the end of her role as an independent agent capable of taking significant initiatives on her own territory. With Unification she became a municipality of the Kingdom like all the others, and she no longer benefited from the daily attention or especially the major projects accomplished by the Venetian aristocracy in the centuries of her greatest splendor. The belated attempt made in 1907 to restore the Tribunal for Waters, a central authority for ongoing repairs to the lagoon, was not successful. The revived institution couldn't find a field for action or the autonomous power for achieving its purposes. The city had to adjust to the small-scale maneuvers of a local ruling class which, though not without thoughtful members and able specialists, was not strong enough to stand apart from the other Italian municipalities; and a national political class that certainly did not surpass it in breadth of vision or historic and cultural awareness.

In fact, for a good part of the 20th century other themes dominated public life in the country and in Venice, while the problems and risks threatening the city remained little known and little discussed. As can easily be imagined, public attitudes concerning the balance between nature and man were very different then, and limited to a few elites.

Perhaps it's no exaggeration or anything new to say that a decisive change in attitudes towards Venice began to appear after the flood of 4 November 1966. The alarm caused by that event marks the beginning, in Venice and in Italy, of a substantial, more lasting and broader attention to the fate of the city, a concerned interest that has steadily grown in the last decades. The resulting discussions and studies found their first substantial legislative expression in law No. 171 of 16 April 1973, concerning "Interventions for safeguarding the Venetian Lagoon," which declared the city of "preëminent national interest." This was the first special law passed for the city. It caused the City in the Lagoon to be numbered among the most significant national concerns, and its preservation an obligation of the State.[1]

As for the legislative and public policy aspects, this was surely a new direction in the history of the city. But most of all, with law No. 798 of 29 November 1984, "Further interventions for safeguarding Venice"—passed unanimously by the Italian Parliament—the outline was drawn for a large-scale strategic intervention that has inspired all the subsequent legislation reinforcing and refining it. The law provides for a Supervisory Commission comprising many ministries and agencies (Merchant Marine, Environment, Cultural Heritage, the mayors of Venice and Chioggia, etc.), and it assigns to a private group, the New Venice Consortium *[Consorzio Venezia Nuova]*, the task of carrying out in the lagoon basin all the interventions coming under State authority, based on decisions made from time to time. So at last a working unit has been identified as responsible for taking on the whole complex of studies, projects and interventions, without which the countless debates and writings on the problems of Venice would just be inert and inconclusive expressions of good will.[2]

The intervention program of the New Venice Consortium—which has recently inspired numerous working agreements with other agencies, mobilizing $84 billion in funding as of December 1994—is planning, and in part has already begun, modification and redesign throughout the range of installations concerning the lagoon.[3] They have been aimed first all, as we shall see briefly, at strengthening the inlets and breakwaters, reinforcing the coastal zone. Of course the lagoon itself has not been foreign to the interests of the consortium. The most significant progress made so far has been scientific, with a thorough and detailed knowledge of events in the lagoon, and identifying the means for intervening with the greatest prospect of success.

Integral to this undertaking is the plan for a complete environmental readjustment of the lagoon, and as can easily be imagined, this requires a fundamental choice in economic and political terms: elimination of oil-tanker traffic from the entire area. This decision and this operation are both high profile, in economic and social as well as logistical terms, and they will surely not be easy to put into practice. They involve balances and powers of general scope, reaching beyond Venice and her immediate hinterland. They involve moving a productive activity that furnishes income and wealth to thousands of families today. But there is no alternative. As gradually as necessary, the City in the Lagoon must surround herself with economies that are not irreconcilable with her survival. She must, in the words of Mayor Massimo Cacciari, "go beyond" the "modern," represented by basic industry, "heavy" transport and pollution, and go back to being a "capital" of research, of the advanced public sector, of specialized services and "intangible" production.[4] In fact, all of Venice will have to rethink her function as a city, to avoid falling, from inertia, into the blighting monoculture of the tourism industry. As has been observed, Venice's urban and environmental heritage must not be degraded to the "rank of an attraction, a market, a stage set available to the highest bidder," but must be developed as a "resource for culture, research and work, so that innovation may be grafted onto the tree of history, planted in the soil of the environment.[5]

3. A Story Just Begun: Rescue

If the study and design phase has been long, dating from the mid-'80s, the working intervention phase is still short. The projects begun in the lagoon are not highly visible so far, but they are spread out in space and are starting to counter the erosion trend that has increasingly prevailed in the last decades. In excavating 38 miles of channels to foster exchange in the waters, the specialists found the material with which to rebuild over 740 acres of mud and salt flats.[1] Today the caissons built to restore the salt flats appear to observers as parts of the ecology, covered with spontaneous vegetation. Until recently they were bare expanses of mud. Now the lagoon has returned them to the landscape.[2]

It's just a drop in the sea, of course. But it's a return to a fruitful logic of the past: such an operation restarts the ancient daily activity by which for centuries the Venetians maintained and managed the delicate balance of the lagoon. The interplay of water and land—digging to make space for the water, and using the mud to recreate land—is resuming, however slowly, and with very different means from those of the past.

Again in the lagoon, with the goal of progressively restoring the morphology of the basin, the first interventions have been defined for the periodic and controlled reopening of the fish pens. Such a change would eventually foster the tidal currents in the lagoon, improve the quality of the surrounding waters, and strengthen the renewable fishing economies that are still being practiced. Restoring the ancient relationships between men and the lagoon is certainly not a small objective: among its goals must always be considered the need to employ individuals once again in the daily maintenance of the waters.

To defend against assaults from the sea, design and construction have advanced. The first construction phase has begun: a grand plan of defense for about 37 miles of shoreline, from the mouth of the Piave to that of the Adige, to withstand storm surges. And the project wisely includes the restoration of the sand dunes that the latest coastal engineering considers precious for absorbing the force of the waves. For that purpose a few miles of dunes have been restored at Cavallino and Pellestrina. "The dune systems," it was recently recorded, "have an important function in defending the land against floods from storm surges by moderating the action of the wind in moving sand, and as a sand reserve for the shore during the most severe surges."[3] Rapid, and in many ways amazing, was the restoration carried out on the other shore, arousing amazed curiosity and keen interest in the public: "The case of Pellestrina was spectacular: from nowhere, in front of the historic murazzi, a wide beach appeared in a short time."[4]

Specialists recognizing the importance of sand is a good sign for the historian. As we saw in the opening pages, the dunes that the Venetians considered precious as early as the 14th century are being rebuilt, with new conservation awareness, by 20th-century engineers.[5]

The protective interventions targeted on habitats have been more extensive. The sea and especially the phenomenon of abnormally high tides are the chief concerns of the specialists.[6] Experimental interventions have been started in some localities (Sottomarina, Pellestrina, San Pietro in Volta, Malamocco, Treporti), building the so-called *insulae* [blocks]. Here the old pavements have been raised and their water sides reinforced. These are new barriers, therefore, ones that can protect dwellings from even exceptionally high tides. For example, the settlement of Malamocco, on the Lido shore, appears to be protected from high tides up to 68 inches. One project includes the construction of new quays and the installation of gates in the canals that cross the settlement. It does not appear possible to do as much, at least for the time being, in centers like Venice and Chioggia because of the unavoidable alteration the architectural and artistic structures in those two urban jewels would suffer.[7] But in recent years, on the initiative of the Cacciari administration, a few experimental interventions of great importance have been made, like raising the "Casa dei Sette Camini" by 16 inches.[8]

To safeguard the whole city, on the other hand, a complex and demanding program has been planned, although it is still the object of controversy and debate: a system of temporary closure by means of mechanical gates to be activated and regulated at times of high tides. The sea enters the lagoon by three inlets, but they cannot be closed permanently because that would cut off the shipping traffic headed for the city. Moreover, without a continual exchange with the sea, the lagoon would stagnate and die in short order, especially in the present phase of acute man-made and industrial pollution. The sea, with its daily contribution of new and oxygen-rich water, is almost solely responsible for the cleaning of the lagoon.[9] The dilemma can be resolved by a bold technical solution: a gigantic arrangement of movable gates to be opened and closed as needed. A full-scale trial of the Experimental Electro-mechanical Module *[Modulo sperimentale elettromeccanico,* or *Mo.S.E.]* has been made with all the components that can't be simulated in miniature. In the summer of 1998 the Mo.S.E. passed the test of environmental compatibility. In terms of planning and technology, it represents a great achievement that can contribute mightily to the future defense of Venice.

Despite the small number of practical results visible today, we must say that this complex of projects and measures is beginning to produce important results for the chances of conserving the city. The Mo.S.E. as well as other projects related to safeguarding the lagoon have been made possible thanks to the tremendous progress made since the '70s in the field of mathematical modeling. These are highly formalized studies oriented towards learning the dynamics of water and the propagation of wave motion in the lagoon, carried out thanks to numerous campaigns of geophysical investigation, depth soundings, and underwater morphology of the basin, exploration of currents and wave motions inside the lagoon, at the inlets, along the shores and near the outer breakwaters.[10]

A contributor in recent years has been the Experimental Center for Hydraulic Models in Voltabarozzo, in the province of Padua, that has reached high levels of specialization in the field. On the technical and scientific levels, then, the results are significant because they have gone beyond purely empirical and conjectural knowledge—even when it comes from centuries of observing the lagoon—reaching an exhaustive scientific framework for the phenomena that can also be simulated and observed in the laboratory.

Several decades of inertia have finally been overcome, and a historic new course has surely begun in comparison to the long silence of the past decades. Today the financial, technical and scientific resources exist to confront the many processes of degradation and destruction that threaten the city; and a new political and cultural awareness has spread among the administrators, the citizens of Venice, and Italian environmental groups. The first four years of the Cacciari administration strengthened the drive of the last decade to put Venice at the center of a concerted, long-term effort on behalf of its survival and the revival of its life as a city. For the first time in forty years the city has a new long-term urban plan. Venice, her development and her transformations are once again the focus of a project collectively conceived and planned for the future. Consonant with that effort the administration has initiated a program of reusing old buildings (e.g. the former convents of Sant' Anna, le Terese, and San Daniele, and the State Storehouse) to obtain over a thousand dwellings in the city and on the islands. This intervention combines with other initiatives aimed both at extending the private restoration of dwelling units and aiding families to buy or rent apartments.[11]

It's a difficult policy, in many respects original, intended to remedy a spontaneous process that is emptying Venice of its inhabitants and thus depriving her of the social forces that must oversee her daily maintenance. Politics is called on here to operate as in a laboratory, experimenting with new rules, or perhaps restoring some of the old standards. Indeed, the resumption of dredging the canals, long abandoned, had a meaning charged with symbolic value, "restorative" of ancient practices of the Republic. Neglected for decades, the canals were full of mud and rubbish of every kind, the source of evil smells that in summer spread to many parts of the city. So an ancient practice of city maintenance was resumed, not without relation to the goal of meeting the need for services on the part of the various populations.

> It's not just removal of the mud from the bottom, as in the past, but a complex operation that includes restoring the quays and the foundations of buildings, improving the drains, and rationalizing services (water, gas, electricity, telephone) as well.[12]

So it would be narrow-minded and wrong not to see in this set of interventions and projects a historic new orientation of political will and knowledge, for saving Venice. As in the days of the Republic, at least some of the conditions are being reproduced throughout the city for dealing with nature's adverse effects (aggravated by the private and destructive interests of men) and the possibility of escaping a destiny that was thought to be inevitable. Thanks also to the new awareness of cultural and environmental assets that has developed in Italy, however late, and the involvement of a broad and influential public opinion, Venice is being seen more and more as a unique monument in human history that must not be lost.

As in the past, as in the golden years of the Republic, Venice cannot be sacrificed to petty private desires, and the overriding interest of her preservation is again becoming part of public awareness and gradually finding expression as political will. Venice must resume telling the tale, broadcasting the proud history of her work in safeguarding, and the countless scattered interventions of recent years. Failure to recognize what's been done, ignoring, or even worse scorning the past, weakens hopes for success and the wager placed on the future. Here again the Republic appears as our teacher in building her own hegemony: every day she drew on her past to build glory and pride for every one of her citizens, encouragement and certainty for her future.

Of course, the road to realizing the projects does not look easy. The time required for carrying out the most ambitious undertakings will necessarily be long, while the city seems ever more defenseless against the unpredictable hazards of extraordinary weather. Decisions and choices will have to be made, and "The ancient political energy . . . that once existed in the city herself"[13] will have to be found again. Today the divisions among the various and sometimes opposing "parties"—unlike the days of the Republic, when in the end a decision was always made—threatens paralysis. Inability to decide is the most corrosive force gnawing away at the foundations of Venice. One of the worst problems besetting her is that she and her lagoon are located in a politically fragile country like Italy. The Environment Minister Giorgio Ruffolo, stated the matter well in 1991, pointing out

> the contradiction often seen in our country between the breadth and complexity of certain problems—requiring coordinated management and long time frames—and agents who are scattered in space and limited in time. Surely the large number of agents and authorities does nothing to simplify the fundamental task of confronting an extremely complex problem like Venice within a coherent framework, with coordinated management and generous

time limits. What we have instead is a multitude of agents, constant changes, interruptions and changes of orientation, of local and national administration. Such fragmentation is not a good omen.[14]

Of course, a steady orientation over time is indispensable for carrying out a great undertaking, and political time is measured by the day. Perhaps the gulf has never been greater, in many respects unbridgeable, between the interests of the political class, with its short-term activities, and the large-scale problems that defy time, relying on results in the future and offering advantages of a collective nature, not individual. Add to this the divergence of opinions, the diversity of outlooks, and the constant questioning of every undertaking. Here is an insidious threat against which the Republic built defenses. As Bernardo Zendrini recalls and historians of Venice have often repeated, at one time the city authorities decided to punish anyone who discussed the problems of the lagoon without having the appropriate technical knowledge. Many citizens joined in the debate among specialists and the arguments in progress with fanciful projects or absurd inventions, and just muddied the waters (so to speak). Luckily for them (and for us) today's chatterboxes are safe from the severe penalties handed down by the Republic. Democracy generously and rightly protects freedom of speech, even for incompetents. But inept democracies don't protect great projects from failure, because they don't protect from the deadly game of chatter and the fragmented interests of individuals.

And yet even in this respect today's political leaders and specialists can find inspiration in certain lessons of the past. We think we have given in these pages the example of a way of working, almost a culture of restorative intervention. It's one that always led the authorities of La Serenissima to make decisions—after long study and many comparisons and debates—and put in practice a physical modification or a new regulation, always watching the results "for the sake of experiment." Always ready, therefore, to revise, correct, and change. Politics attentive to the lessons of the facts: that was the truly modern way of working the Venetian authorities inaugurated for the lagoon, centuries ago.

And yet we must recognize there are problems that outstrip the abilities of Venetian and Italian leaders to make decisions and choices. Not everything will be decided in this small corner of the planet. The rise in sea levels, reported in several places and perhaps already visible, threatens the island today. If it is not stopped, it will finally doom

Venice—in spite of all local attempts at defense, and once again in some indefinite future—to sink into the waters of the Adriatic.

The destiny of the city in the lagoon appears closely linked to worldwide events, much more so than in the past, and the global nature of environmental events, which is now obvious in our time, appears in the case of Venice with dazzling immediacy. Today Venice "belongs" to the world in this new and worrisome sense as well. If global warming, due to productive activities and poisonous emissions on a colossal scale, keeps on melting the arctic ice, the seas will keep relentlessly rising.[15] Meanwhile Venice is there, waiting in the gulf of the Adriatic, and her destiny depends in part on the destiny of the planet itself.

So if her centuries of history give us a unique example of economic growth combined with preserving environmental balance, if the experience of the Republic shows us an unattainable model of political farsightedness and governmental wisdom, then its singular case is, doubly, a "model for our planet." Venice shows us and indeed exemplifies, in her present condition and on a city-wide scale, the outlook for our planet. At the same time, her past experience and the success of her policy show us the chance to face the huge challenges of the future with some possibility of success.

Afterword, 1995

Massimo Cacciari

When examining the problems facing Venice, discussion often neglects the achievements and the modifications made by La Serenissima over the centuries. We often find that those achievements and those modifications are not generally known, and historical references are often limited to emphasizing how certain problems have endured; they show no clear understanding of the innovative and outstanding ways in which the Venetian Republic dealt with the issue of safeguarding the city and her lagoon.

For these reasons, the present book is very useful: it carefully documents the fundamental issues and compares the various options, setting out the reasons for the choices that were made and describing the actions carried out by the Most Serene Republic.

Those actions were always based on a full awareness, not entirely present in our time, that the "end" of the city and its lagoon might result not just from one or more catastrophic and unforeseeable events, but from human intervention (or failure to intervene) as well.

How can we fail to use the same caution the Venetians showed when undertaking changes to the lagoon, or their wisdom in making adequate experiments beforehand, or again their conviction that in every case the safety of the city lies entirely in the hands of those who live in it? In this connection, the author rightly emphasizes the fact that unlike any other example from that time (and not excluding examples yet to come), the power of the ruling class in the Most Serene Republic was legitimated solely by their ability to guarantee the survival of the city and the preservation of the lagoon. What Venice succeeded in achieving in past centuries can be defined as wise "natural-resource policy."

The Venetians had a "special aptitude for economy" because they had a precise and correct sense of the word "economy." That's the only explanation for the body of laws, justly cited for its "stability and vigor," that consistently controlled the use of resources as a function not only of their necessary renewal, but also their proper use in the development of productive activities. In addition, Venice's laws and her magistrates "vigorously and severely" defended the lagoon against all those who intended to exploit it without regard for preserving balance in the waters and the environment. These are the main reasons why the Most Serene Republic was able to resolve successfully the issue of the city's survival.

Regarding the procedures, regulations and modifications just described, the Venetian ruling classes managed to secure a broad "mobilization of consensus" such as never could develop from mere "deregulation," as we say today, or on the other hand from a policy consisting solely of limits on freedom of action for private individuals.

All this is reflected even in the last acts of the Serenissima, when she was already being overwhelmed by history: she marked out the "lagoon boundary," a line that once again defined the State's ownership of the waters; and she undertook the colossal project of the murazzi to protect the seashore. These acts were not "random"; they showed then and they still show, to governments that have succeeded La Serenissima, the high road to follow. So we may conclude, as the author rightly does, that in that fateful year of 1797 "Venice had mastered her destiny" and reached her "ultimate goal" of survival. But today, almost two centuries after the Republic fell, can the Venetians say as much? The answer is all too obvious and can only be negative, but rather than dwell on the countless reasons for it, we must briefly review the opportunities Venice still has for becoming mistress of her destiny once again.

The crux of the issue is still "resources": natural, economic, and social. Resources that the City in the Lagoon still possesses in spite of everything, representing the basis for her longed-for revival. For some time yet to come Venice can "exploit" her image as the center of an absolutely unique historical and environmental context, but she must do so wisely—and here lies the connection to the experience of La Serenissima—in a completely new and different way from the past decades. Tourism, the masses of visitors that threaten to bring about the "end" of the city, are neither a natural force nor an unforeseeable one, and they must be transformed into a decisive factor for revival. The waves of tourists must be "channeled," linked closely to the reorganization of cultural activities in Venice, the museums and the creation of the "scattered museum," and the careful, intelligent use of the incomparable lagoon environment.

Venice is still a place of transit, and it will become more and more so in the context of Europe, especially Eastern Europe and its inevitable development after the terrible time of the World Wars. Therefore, the port and the airport have become the leading vehicles for the economic revival of the lagoon region.

This historic city is once again the ideal base for activities based on scientific research, and the ideal site for technological research, at a time when the immaterial dominates—information networks that span the globe—as a modern continuation of the activities of the Arsenale in the days of La Serenissima.

After all, there is no contradiction between grand strategic designs and the daily conservation and maintenance of the city and the lagoon, both critically urgent tasks (think of the condition of the city's canals, or the distressed condition of the lagoon), and an exceptional opportunity for developing the latest techniques for structural and environmental restoration, in keeping with the great tradition of Venice's golden age.

These are great opportunities that we absolutely must manage to seize, for they bear witness to a single great possibility: Venice can become a place for investing significant human and economic resources, a place where "work" can resume with confidence. In order to "take charge of its destiny," finally, the City in the Lagoon needs leaders who are equal to the task. "Leaders" as distinct from a "political establishment," because Venice needs great entrepreneurs, great leaders who can seize opportunities and turn them to account, leaders who are able to act, and act well, in concrete terms, according to realistic projects that are financially responsible and in every way respectful of the delicate fabric of the city.

Venice
January 1995

Note to the 1998 Edition

This new edition of Piero Bevilacqua's book could not come at a more appropriate time. In the immediate future questions of the greatest consequence for the saving of Venice will have to be decided. In particular, the decisive debate is beginning on the project of moveable gates for the port inlets, whose importance Bevilacqua emphasizes. The new chapters of his book summarize very clearly all the problems on the table. His work is particularly valuable because he does not indulge in laments or dwell on things not done, on the delays or the imminent disasters, or on the usual arguments on behalf of the quality and effectiveness of the projects set forth. The value of Bevilacqua's book was and still is that it explains the exceptional complexity of the problem that is Venice. His analysis clearly sets forth the priorities for the task of rescue. I shall summarize them here, indicating in each case the present status of the issue.

1. Ongoing erosion requires gigantic restoration projects, to prevent the destruction of the Lagoon's very nature. In this regard the delay is obvious.

2. Pollution abatement is a precondition for any and all projects at the port inlets; unless reduced, pollution could cause the situation to deteriorate to an unacceptable degree.

3. Industrial activity at Marghera must or can be cleaned up. In this case we are one step away from a permanent agreement between Enichem, the Ministry of the Environment and the Municipality.

4. Reinforcing the seaward jetties and the rest of the coastal protection is an absolutely strategic task. It is already well advanced and will surely be completed within a few years.

5. All the tasks that are only mentioned in the book, those directly concerning the historic center of Venice—dredging canals, raising parts of the lowest areas of the City, reinforcing foundations, and improving the sewer system—are already under way. Dozens of projects are now

in progress, and for dredging canals alone the expenditure is about $40,000,000 a year.

6. Together with these projects, an extraordinary plan is going forward with restoration work on the monuments and museums of the City, as well as private buildings. Plainly, the situation is no longer a jumble of discordant projects, occasional, fugitive and superficial, such as those which defined the Venetian Lagoon until fairly recently.

Naturally, such a huge volume of projects will not be possible without the aid of the national and international communities. I hope that Piero Bevilacqua's book, in this new version, may help us to move in that direction, showing the reality of Venice today for what it is, shadows and light.

Venice M.C.
September 1998

Notes

Abbreviations

ASV	Archivio di Stato di Venezia
AZ	Secreta Archivi Propri. Archivi Zendrini
BNM	Biblioteca Nazionale Marciana
GV	Giustizia Vecchia
PsB	Provveditori sopra i Boschi
PBI	Provveditori sopra i Beni Inculti
SEA	Savi ed Esecutori delle Acque

Chapter 1, Section 2 – The Lagoon: A Natural Bastion

1. *Memorie storiche dello stato antico e moderno delle lagune di Venezia e di que' fiumi che restarono divertiti per la conservazione delle medesime di Bernardino Zendrini, matematico della Repubblica di Venezia*, Venice, Stamperia del Seminario, 1811, vol. I, p. 1.

2. Giacomo Filiasi, *Osservazioni sulle cause che possono aver fatto ritrovare nel secolo XIV in parte pregiudicata la Laguna rispetto alla posizione di Venezia*, Opuscoli due, Venice, Francesco Andreola, 1820, p. 10. By the same author, see also *Memorie storiche de' Veneti primi e secondi*, Venice, 1796, vol. I, pp. 16ff. Another author recalls that the mainland portion of the Domain lying between the Po and the Isonzo "carries within it more than six hundred rivers and streams" (Cristofaro Tentori, *Saggio sulla storia civile, politico, ecclesiastica e sulla corografia e topografia degli stati della Repubblica di Venezia*, Venice, Giacomo Storti, 1785, vol. II, p. 215).

3. Gio Malaspina, *Memoria storico-idraulica sul porto di Lido*, Venice, Stabilimento tipografico Grimaldo & C., 1871, p. 36.

4. *Considerazioni intorno alla Laguna di Venezia di Fra Stefano Angeli Matematico dello studio di Padova*, ASV, SEA, AZ, b. 20 (29 July 1665).

5. Elia Lombardini, *Studi idrologici e storici sopra il grande estuario adriatico e principalmente gli ultimi tronchi del Po*, Milan, G. Bernardoni, 1869, p. 9. See also Paolo Morachiello, *Le bocche lagunari*, in *Storia di Venezia*, XII, *Il mare*, eds. A. Tenenti & U. Tucci, Rome, Istituto dell'Enciclopedia Italiana, 1991, pp. 77-8. The process of lagoon formation was the same as the one which formed salt lakes along the Southern Italian coasts, described in the 19th century by Carlo Afan de Rivera: cf. Piero Bevilacqua & Manlio Rossi-Doria, *Le bonifiche in Italia dal 700 a oggi*, Bari & Rome, Laterza, 1984, p. 43, note 98.

6. The names listed above are those of the eighteenth-century ports, also cited by Tentori, *Della legislazione veneziana sulla preservazione della Laguna. Dissertazione storica-filosofica-critica del Sig. Abate Cristofaro Tentori*, Venice, Giuseppe Rosa, 1792, pp. 156-7. But the seaward passes have changed over time, in name and number, and in earlier times it appears there were indeed seven, before they were reduced to today's three actual passes. Cf. Erminio Cucchini, *La laguna di Venezia e i suoi porti*, Rome, Stabilimento tipo-

litografico del Genio civile, 1912, p. 31.

7. For example, in a measure adopted 7 September 1334, the Senate prohibited the removal of sand from the shores (Tentori, *Della legislazione veneziana*, op. cit., pp. 228-9). For a more general treatment, see Maria Francesca Tiepolo, *Difese a mare*, in *Mostra storica della laguna di Venezia*, Venice, Stamperia di Venezia, 1970, p. 33.

8. Giacomo Filiasi, *Memoria delle procelle che annualmente sogliono regnare nelle maremme veneziane*, Venice, Antonio Zatta e Figli, 1794, pp. 13-14. Of course, other kinds of storms were just as harmful. See Filiasi, *Memorie storiche de' Veneti,* vol. III, p. 7. A major 19th-century scientific inventory is found in *Carta topografica idrografica militare della Laguna di Venezia, ordinata da sua altezza imperiale Eugenio Napoleone vice re d'Italia eseguita [... da] Augusto Denaix* [1809-1811], in Delegazione italiana della Commissione per l'esplorazione scientifica del Mediterraneo, *La Laguna di Venezia*, monografia coordinata da Giovanni Magrini, Atlante terzo, Venice, Officine Grafiche C. Ferrari, 1933. More recent is Giuseppe Crestani, *Il clima*, in Delegazione italiana della Commissione per l'esplorazione scientifica del Mediterraneo, *La laguna di Venezia*, ed. G. Mangrini, Venice, Officine Grafiche C. Ferrari, 1933, vol. I, part II, tome III, pp. 220ff.

9. Alvise Da Mula, in *Relazioni dei Rettori veneti nel Dogado. Podestaria di Chioggia*, ed. B. Polese, Milan, Giuffrè, 1982, p. 21.

10. *Relazioni periti area la Laguna (1725-1734),* ASV, SEA, b. 55. The document, dated 9 July 1725, is a petition from a boatman charged with transporting 200 barges of mud dredged from the Mazzorbo channel. This was, plainly, an old and recurrent problem: on 8 July 1335 the Senate ordered the owners of vineyards on the Lido of Malamocco to reinforce the levees and prevent the collapse of mud and earth into the Lagoon and the sea (Tentori, *Della legislazione veneziana*, p. 229).

11. Luigi Lanfranchi e Gian Giacomo Zille, *La laguna*, in *Storia di Venezia, II, Dalle origini del Ducato alla IV Crociata*, Venice, Officine Grafiche C. Ferrari, 1958, p. 43; Bianca & Luigi Lanfranchi, *La laguna dal secolo VI al XIV,* in *Mostra storica della laguna di Venezia* (1970), p. 80; and Jean-Claude Hocquet, *Expansion, crises et déclin des salines dans la lagune de Venise au Moyen Âge*, ibid., p. 92.

12. For the working of the wells, see Massimo Costantini, *L'approvvigionamento idrico della Serenissima*, Venice, Arsenale, 1984, p. 19. For the late Middle Ages, see Elisabeth Crouzet-Pavan, *«Sopra le acque salse»: Espace, pouvoir, et société à Venise à la fin du Moyen Âge*, Rome, Istituto storico italiano per il Medioevo, 1992, I, pp. 244ff. The quote refers to Piero di Fanti, *Che le acque bianche vadino per li suoi torrenti et non atterrerano la Laguna* (Let the white waters run in their beds and they will not fill in the Lagoon), in R. Cessi e N. Spada, eds., *Antichi scrittori d'idraulica veneta*, in *La difesa idraulica della laguna veneta nel sec. XVI. Relazioni dei periti*, Venice, Officine Grafiche C. Ferrari, 1952, p. 25.

13. Filippo di Zorzi, *Dell'aria et sue qualità. Discorso*, Venice, Gio. Ant. Rampazetto, 1596, p. 8.

14. *Scrittura presentata al Senato Veneto il dì 18 Giugno 1740*, in *Documenti autentici che dimostrano doversi unicamente a Bernardino Zendrini [...] il progetto per la costruzione del celebre riparo detto i murazzi dei lidi di Venezia*, ed. Angelo Zendrini, Venice, Tipografia Alvisopoli, 1835, p. 15. In general the persons thought responsible for such acts were the very ones employed building the palisades.

15. *Suppliche e scritture al Senato* (undated writings), ASV, SEA, b. 268. This document probably dates from the second half of the 16th century. On the protection of the *lidi*, see Eugenio Miozzi, *Venezia nei secoli*, in *La laguna*, Venice, Il Libeccio, 1968, pp. 167ff.

16. Tentori, *Saggio sulla storia civile*, Vol. II, p. 213. The quote is from a medical doctor, Andrea Marini, *Discorso sopra l'aere di Venezia [...]*, in *Antichi scrittori di idraulica veneta*, ed. A. Segarizzi, Venice, Tipografia Alvisopoli, 1923, p. 5.

17. *Relazione di Domenico Guglielmini del 17 febbraio 1699,* ASV, SEA, AZ, b. 20.

Chapter 1, Section 3 – The Silent Enemy: Silting

1. *Discorso sopra l'origine delle atterrazioni della Laguna Veneto antica e moderna [...] del dottor Carlo Antonio Bertelli Accademico Simpatico, Pacifico e Disinvolto* [sic], Venice, A. Bosio, 1626, p. 17.

2. Girolamo Fracastoro, *Lettera sulla Laguna di Venezia (sec. XVI) ora per la prima volta pubblicata ed illustrata [...]*,Venice, Tipografia Alvisopoli, 1815, pp. 9-10.

3. Ministero dei Beni culturali e ambientali, Archivio di Stato di Venezia, Catalogue of the exhibition, *Ambiente e risorse nella politica veneziana*, Venice, 5 August – 8 October 1989, p. 40.

4. See Cessi, *Evoluzione storica del problema lagunare*, Venice, Officine Grafiche C. Ferrari, 1960, p. 21, and Paolo Rosa Salva & Sergio Sartori, *Laguna e pesca. Storia, tradizioni e prospettive*, Venice, Arsenale, 1979, p. 13.

5. *Della laguna*, in *Antichi scrittori d'idraulica veneta, I, Marco Cornaro (1412-1464), Scritture sulla laguna*, ed. G. Pavanello, Venice, Officine Grafiche Ferraresi, 1919, pp. 81-2. Zendrini also gives contemporary accounts of the unhealthy conditions caused by the incident, *Memorie storiche*, II, pp. 96-7.

6. *Istruzioni de messer Christofaro Sabbadin Dal Friol circa questa Laguna et come l'era anticamente, et come la si trova al presente, et le cause della ruina de quella, con il modo de salvarla et farla eterna, date al chiarissimo messer Vicentio Grimano, Procurator et Savio sopra le acque, in 1540 del mese di febraro*, in Cristofaro Sabbadino, *Discorsi sopra la laguna*, in *Antichi scrittori d'idraulica veneta*, ed. R. Cessi, Venice, 1930, vol. II, part I, p. 10. Concern that Venice might become unlivable in the short or long term must have been considerable if the Council of Ten, in a 1501 document about instituting the Sages, defined their purpose as combating "so incomparable an ill, that alone can make this city of ours and our most loving [sic] and dear land uninhabitable" [Latin] (Paolo Selmi, *Politica lagunare della veneta repubblica dal XIV al XVIII secolo*, in *Mostra storica*, p. 109).

7. Marini, *Discorso sopra l'aere di Venezia*, p. 5.

8. Benedetto Castelli, *Considerazioni intorno alla laguna di Venezia [1641]*, in *Raccolta d'Autori che trattano del moto dell'acque*, Florence, Stamperia di Sua Altezza Reale, 1765, vol. I, p. 151.

9. *Scritture e proposizioni per migliorare lo stato della Laguna,* ASV, SEA, 1672-1673, b. 128.

10. However, it appears that over time the cuts produced positive results. Bernardo Zendrini gave a favorable opinion; a century later, he was able to see improvements

in the Lagoon as well as in the port of Malamocco. See Zendrini, *Memorie storiche*, II, p. 72.

11. *Considerazione intorno alla Laguna di Venezia di Fra Stefano Angeli*, ASV, SEA, AZ, b. 20.

12. On the connection between building levees around the fish pens and the increase of malaria in the Lagoon, see Rosa Salva & Sartori, *Laguna e pesca*, p. 13.

13. Document dated 28 July 1536, ASV, SEA, Capitolare (1530-1538), b. 343. On waste water from dye works, see Crouzet-Pavan, *«Sopra le acque salse»*, I, p. 296; already in the Middle Ages she sees signs «d'une ville en danger» (p. 314). Prohibitions of these abuses are documented starting at the end of the 13th century at least, «lest water or anything foul fall into the canals, rios, basins, or waterways outide the walls» [Latin], *Laguna, lidi, fiumi. Cinque secoli di gestione delle acque,* exhibition, 10 June – 2 October 1983, curated by Maria Francesca Tiepolo, Venice, Tip. Helvetia, n.d., p. 20. There were places in the City designated for depositing garbage—the *Caselle delle Scoazze*—maintained by the *Nettadori* (cleaners) of each Sestriere (Tentori, *Della legislazione veneziana*, p. 215). The garbage was then transported in special boats (*burchi*) as fertilizer for gardens: cf. Gianpietro Zucchetta, *Una fognatura per Venezia. Storia di due secoli di progetti*, Venice, Istituto veneto di Scienze Lettere e Arti, 1986, pp. 11-12.

14. *Relazioni dei Periti e Disposizioni 1583-1591. Relazione del Proto de' Lidi del 22 agosto 1583*, ASV, SEA, b. 158.

15. Tentori, *Della legislazione veneziana*, p. 277.

Chapter 1, Section 4 – Navigating the Lagoon

1. ASV, SEA, Capitolare (1530-1538), b. 343. For difficulties in the ports, see C. F. Lane, *Storia di Venezia*, Turin, Einaudi, 1991, p. 22. An important sign of the navigation problems caused by silting in the 14th century is the order dated 13 July 1333 whereby the authorities required all vessels to take on ballast dredged from a heavily silted location (Punta di Santa Maria), thus improving navigability without burdening the treasury (Tentori, *Della legislazione veneziana*, p. 199).

2. *Relationi de NN.HH. Esecutori et itinerari per visite alla Laguna, 1547-1670*, ASV, SEA, b.142.

3. Loc. cit.

4. Ibid., Report dated 18 August 1574.

5. Ibid., Report dated 2 February 1580.

6. Ibid., Report dated 20 March 1581.

7. [N. Contarini], *Opera profittevole, e necessaria causata dall'autore dall'esperienza d'una lunga pratica per migliorar e conservar la Laguna di Venetia*, Milan, Ambrogio Ramellati, 1675, pp. 24, 27.

8. See for example Bernardo Trevisano, *Della Laguna di Venezia. Trattato*, Venice, Domenico Lovisio, 1718, p. 77, and Carlo Silvestri, *Istorica e geografica descrizione delle antiche paludi adriane ora chiamate lagune di Venezia [...]*, Venice, Domenico Occhi, 1736 (facsimile edition, Bologna, Forni, 1973), p. 202.

9. *Relazioni periti circa la laguna* (1725-1734), ASV, SEA, b. 55, and SEA, b. 56.

10. *Scrittura nell'universale della materia di Laguna, con li riflessi al suo Interno, e a' Fiumi, scoli, lidi e porti del Magistrato Eccellentissimo all'Acque servito dal Fiscale Giulio Rompiasio*, Venice, 1714, p. 2. We must not overlook the size of certain ships, such as the *caracche* and *galere grosse*, "which are really seagoing castles" in the opinion of a contemporary (Antonio Donati, *Trattato de semplici, pietre et pesci marini che nascono nel lito di Venetia, la maggior parte non conosciuta da Teofrasto, Dioscoride, Plinio, Galena, et altri Scrittori*, Venice, Pietro Maria Bertano, 1631, p. 8). A contemporary exception was Shakespeare, who must have seen on the Thames ships that "Like signiors and rich burghers on the flood,/Or, as it were, the pageants of the sea,/Do overpeer the petty traffickers" [*Merchant of Venice*, I, i] (quoted in Lane, *Storia di Venezia*, p. 440).

11. *Scrittura nell'universale*, p. 7.

12. During the past forty years, as Zendrini recalled on 18 April 1726, the port of Malamocco "may be considered one of the best in all Europe." *Relazioni periti circa la Laguna* (1725-1734), ASV, SEA, and Contarini, *Opera profittevole*, pp. 11ff.

13. Cf. Malaspina, *Memoria storico-idraulica sul porto di Lido*, pp. 35ff., and Miozzi, *Venezia nei secoli*, III, *La laguna*, pp. 153ff.

14. Francesco Calcaneis, *Scritture intorno alla laguna*, n.d., p. 9. The document containing the quote is dated 12 February 1692. Similarly, and indeed like other writers, Tentori saw as subordinate to the condition of the Lagoon "the four chief assets of the Ruling City: health, which is liberty, security, and trade" (*Saggio sulla storia civile*, vol. VIII [1787], p. 222).

15. Predag Matvejevic', *Golfo di Venezia*, Milan, F. Motta, 1995, p. 10.

16. Marco Bandesan, *L'evoluzione geologica del territorio veneziano*, in *Mostra storica*, pp. 37-8, Paolo Selmi, *Politica lagunare della veneta repubblica dal secolo XIV al XVIII*, ibid., p. 105, and Crouzet-Pavan, *«Sopra le acque salse»*, I, pp. 336-7.

17. Filiasi, *Osservazioni sulle cause*, p. 128.

18. *Relazione de' Periti e Disposizioni* (1583-1592), ASV, SEA, b. 158. Testimony of Cristofaro Sabbadino, 25 August 1576.

19. *Deposizione de' Periti et altri circa la Laguna dal 1493 sino al 1579*, ASV, SEA, AZ, b. 20, Report dated 8 June 1551, containing the testimony of several fishermen of San Nicolò.

20. Ibid. p. 108.

Chapter 1, Section 5 – Present Dilemmas and Future Mystery

1. *Opera profittevole*, pp. 17-18.

2. *Memorie storiche*, II, p. 72.

3. *Scritture e proposizioni per migliorar lo stato della Laguna* (1672-1673), ASV, SEA, b. 128.

4. *Relazioni periti circa la Laguna* (1725-1734), ASV, SFA, b. 55. This is a report on the condition of the existing fish-pens *(valli)* in the Lagoon dated 18 April 1726.

5. *Discorso sopra l'aere di Venezia*, p. 29.

6. See *Raccolta d'autori che trattano*, p. 163.

7. Simon Schama, *La cultura olandese del secolo d'oro*, Milan, Il Saggiatore, 1988, p. 25.

8. Ibid., p. 70. The Venetian elite, as shown by a study of individual personalities, was profoundly attached to history as a form of knowledge particularly useful in facing the problems of the present. Cf. Peter Burke, *Venezia e Amsterdam. Una storia comparata delle élites del XVII secolo*, Bologna, Transeuropa, 1988, pp. 108-9. But for these aspects of Medieval Venice, see Crouzet-Pavan, *«Sopra le acque salse»*, I, pp. 64ff. An important contribution to the comparative history of these two communities, regarding water and reclamation policies, is Salvatore Ciriacono, *Acque e agricoltura. Venezia, l'Olanda e la bonifica europea in età moderna*, Milan, Franco Angeli, 1994.

Chapter 2, Section 1 – Fresh Water and Brackish Water

1. G. Beloch, *La popolazione di Venezia nei secoli XVI e XVII*, in «Nuovo Archivio Veneto», 1902, III, Part I, p. 48.

2. Cf. Daniele Beltrami, *Storia della popolazione di Venezia dalla fine del XV secolo alla caduta della Repubblica*, Padua, Cedam, 1954, pp. 59-71. Venice's population must have grown rapidly in the previous centuries as well, especially between the Fourth Crusade (1202-1247) and the Black Death of 1348. See Gino Luzzatto, *Storia economica di Venezia dall'XI al XVI secolo*, Venice, Officine Grafiche C. Ferrari, 1961, p. 38.

3. For all these details see Costantini, *L'approvvigionamento idrico.* But already in the 14th century the authorities occasionally bought water to refill these wells in dry times: cf. Crouzet-Pavan, *«Sopra le acque salse»*, I, p. 250.

4. Costantini, *L'approvvigionamento idrico*, p. 16.

5. *Expansion, crises et déclin des salines*, in *Mostra storica*, p. 87.

6. Ibid., *Le Saline*, in *Storia di Venezia,* I, *Origini—età ducale.* Rome, Istituto dell'Enciclopedia Italiana, 1992, p. 520.

7. On the commercial aspects of this product and City's pivotal role in trade over the years, see the monumental work of Jean-Claude Hocquet, *Il sale e la fortuna di Venezia*, Rome, Jouvence, 1990.

8. Filiasi maintained that the salt pans "ringed with levees and walls, making of them so many low islands, may well have created obstacles to the free ebb and flow of the tide." Together with concern for healthy air, this probably led to their gradual removal. Cf. Filiasi, *Osservazioni sulle cause*, p. 87. As late as the mid-16th century salt was occasionally made in the Lagoon of Chioggia. Cf. Polese, *Relazione dei Rettori veneti nel Dogado*, p. 6 (This report, by Rector Francesco Tagliapietra, is dated 1559).

9. Emanuela Casti Moreschi, *L'analyse historique de l'utilisation des eaux dans la lagune de Venise*, in *L'eau et les hommes en Méditerranée*, ed. André de Réparaz, Editions du Centre National de la Recherche Scientifique, Paris, 1987, p. 78.

10. Agostino Sagredo, *Sui mulini che esistevano anticamente nelle lagune di Venezia. Nota storica,* Padua, Randi, 1860, p. 257. "There was no monastery," abbé Tentori recalls, "that did not have its mills" (*Della legislazione veneziana*, p. 112).

11. Sagredo, *Sui mulini che esistevano*, p. 261. Cf. Filiasi's information and opinions in *Osservazioni sulle cause*, pp. 78-9. Another writer convinced that the mills harmed the Lagoon was C. A. Bertelli, *Discorso sopra l'origine delle Atterrazioni*, p. 10. One can get an idea of the extraordinary spread of mills along the course of the Adige by consulting the

Registro Privilegi Molini sull'Adige (1769-1800), ASV, SEA.

12. *Suppliche et Risposta (1574-1581)*, ASV, SEA, b. 270.

Chapter 2, Section 2 – Hunting and Fishing

1. Hannelore Zug Tucci, *Pesca e caccia in laguna*, in *Storia di Venezia*, I, p. 491.

2. G. S. Bullo Giustiniani, *Piscicultura marina. Stima delle coltivazioni in acqua salsa*, Padua, Stabilimento Prosperini, 1891, pp. 246ff. and 390ff., and Giancarlo Ligabue, Pietro Basaglia, & Gabriele Rossi Osmida, *Pesca e caccia nell'antica ecologia lagunare*, in *Mostra storica*, p. 163. See also the various contributions in *Laguna: Conservazione di un ecosistema*, Venice, Comune di Venezia-WWF, 1984.

3. *Proclama publicato de Ordine dell'Illustrissimi Signori Giustizieri Vecchi, 25 Ottobre 1655*, ASV, Stampe Giustizia Vecchia, b. C.

4. Ligabue, Basaglia, & Rossi Osmida, *Pesca e caccia*, p. 164.

5. Government control of sales must be very ancient: in 1173 a law set maximum prices for the various fish. Cfr. Zug Tucci, *Pesca e caccia*, in *Storia di Venezia*, pp. 499-500.

6. For these details see the wealth of information and documents in the publication by the Provincial Government of Venice, *La pesca nella laguna di Venezia. Antologia storica di testi sulla pesca nella laguna, la sua legislazione il suo popolo la lingua e il lavoro dei pescatori, sui pesci e sulla cucina*, Venice, 1989, pp. 7ff. But Venice's whole social life tended to be organized around associations, called *scuole*. There were *scuole* for artisans, as there were for exiles and refugees, for the blind (*Orbi*), and the crippled (*Zotti*). See Brian Pullan, *La politica sociale della Repubblica di Venezia, 1500-1620*, I, *Le scuole grandi, l'assistenza e le leggi sui poveri*, Rome, Il Veltro, 1982, p. 41. The associative aspect of society, and the role of associations in maintaining consensus between government and the governed, are emphasized in Lane, *Storia di Venezia*, pp. 122ff. and 318-9, and Edward Muir, *Il rituale civico a Venezia nel Rinascimento*, Rome, Il Veltro, 1984, p. 41.

7. *Ordini presi nell'Eccellentissimo Collegio delle Pescarie*, 1595 Adì 15 Decembre, Stampati per A. Pinelli Stampator Ducale, BNM.

8. *Proclama dei Proveditori sopra la giustizia vecchia, ed inquisitori ai viveri del 5 aprile 1797*, ASV, Stampe Giustizia Vecchia, b. C.

9. Judgment dated 12 July 1680, ASV, Documenti Giustizia Vecchia, b. C.

10. *Proclama pubblicato d'Ordine degl'Illustrissimi ed Eccellentissimi Signori Proveditori sopra la Giustizia Vecchia, Inquisitori sopra i Viveri, e Giustizieri Vecchi. Collegio Delegato. In proposito delle Limitazioni e Tariffe del Pesce*, 14 June 1752, ASV, Stampe Giustizia Vecchia, b. C.

Chapter 2, Section 3 – "Planting" Fish

1. *La pesca nella laguna*, p. 41, Ligabue, Basaglia & Rossi Osmida, *Pesca e caccia*, in *Mostra storica*, pp. 160-1, and G. B. Rubin de Cervin Albrizzi, *Imbarcazioni lagunari*, ibid., pp. 149-53. On the evolution of Venetian boats designed for navigating the rivers, the Lagoon, and the open sea, cf. Tentori, *Saggio sulla storia civile*, vol. I, pp. 326ff.

2. *Ordeni et capitoli presi nel Collegio detti Clarissimi Signori Proveditori e Giustizieri Vecchi in Materia di ogni sorte di Rede, Spesse, Chiuse, Ostreghere e altro. Stampati per A. Pinelli infine 1599*

Adì 11 ottobre, Publicati sopra le Scale di S. Marco e di Rialto, BNM. The records of a trial for taking young fry, opened 20 April 1761, together with samples of the impounded nets, are in ASV, Giustizia Vecchia, b. 84.

3. *Ordeni et Capitoli presi*, BNM.

4. Magistrati alla GV, ASV, serie I, Capitolari, b. 1.

5. *Ordini presi nell'Eccellentissimo Collegio delle Pescarie, 1595 Adì 15 Decembre*, BNM.

6. *Ordini in proposito della Pesca del Pesce Novello estesesi dagli Illustrissimi ed Eccellentissimi signori sopra la Giustizia Vecchia ed Inquisitori sopra i Viveri approvati con Decreto dell'Eccellentissimo Senato il 29 settembre 1775*, ASV, Stampe Giustizia Vecchia. One of the oldest records showing the Venetian government's concern with protecting fish in the Lagoon dates back to 8 March 1314, when it was prohibited to fish for gray mullet (*cefalo*) before St. Peter's day (29 June). Cf. Giovanni Mazier, *Brevi cenni sulla pesca nella veneta laguna*, Venice, Tipografia Antonelli, 1893, p. 7. In the same year a document prohibits catching or selling "pisces vaninos" (young fry) until St. Peter's day (Zug Tucci, *Pesca e caccia*, in *Storia di Venezia*, p. 493).

7. *Ordini in proposito della Pesca del pesce novello estesesi dagli Illustrissimi ed Eccellentissimi signori sopra la Giustizia Vecchia ed Inquisitori sopra i Viveri approvati con li Decreti dell'eccellentissimo Senato 13 settembre e 6 ottobre 1781*, ASV, Stampe Giustizia Vecchia, b. e, 1781.

8. Judgment dated 18 February 1781, ASV, Stampe Giustizia Vecchia, b. C.

Chapter 2, Section 4 – Hatcheries in the Lagoon: The "Fish Pens"

1. In the 16th century, according to Sabbadino, there were about 61 enclosed and open fish pens *(valli)*, while 62 were listed in 1662. See Rosa Salva and Sartori, *Laguna e pesca*, pp. 11-12. In the first two decades of the 18th century they were reduced to 19, eight in the upper Lagoon and 11 in the lower (*Relazioni periti circa la Laguna* (1725-1734), ASV, SEA, b. 55, *Relazione delle commissioni incaricate di riferire sullo stato delle valli esistenti in Laguna del 24 luglio 1728*). In the middle of the 19th century their number was reckoned at 42, while according to recent surveys there are 24, totaling 21,080 acres (Rosa Salva e Sartori, *Laguna e pesca*, p. 11). For fish pens, on which there is of course a huge bibliography both old and new, see also the detailed reconstruction, based mainly on maps, by Eugenia Bevilacqua, *Diffusion des techniques de l'eau du monde antique au monde contemporain: les «valli da pesca» dans la lagune de Venise*, in *L'eau et les hommes en Méditerranée*, pp. 67ff.

2. *Ordini in proposito della Pesca del pesce novello*, ASV, Stampe Giustizia Vecchia, b. C, 1781. As for the *oradelle* "to be cast into the pens," the Justices ruled that "all pen operators intending to fish their own pens must make themselves known to the Magistrates in February." See also, in this regard, Ligabue, Basaglia, Rossi Osmida, *Pesca e caccia*, p. 163.

3. Zug Tucci, *Pesca e caccia*, p. 495.

4. A. Targioni Tozzetti, *Allevamento degli animali acquatici*, Enciclopedia Agraria Italiana, Turin, Unione Tipografica Editrice, 1880-1882, and at greater length R. Del Rosso, *Pesche e peschiere antiche e moderne nell'Etruria marittima*, Florence, Osvaldo Paggi, 1905. Cf. also Bullo Giustiniani, *Le valli salse da pesca*, in *La laguna di Venezia*, Delegazione italiana della Commissione per l'esplorazione scientifica del Mediterraneo, vol. III, part VI, tome XI, pp. 29ff.

5. See Bullo Giustiniani, *Piscicultura marina*, p. 245.

6. For the concept of resource, see Piero Bevilacqua, *Uomini, lavoro, risorse*, in *Lezioni sull'Italia repubblicana*, Rome, Donzelli, 1994, pp. 113-7.

Chapter 2, Section 5 – An Appetite for Wood

1. *Parte presa nell'Eccellentissimo Collegio delle Legne Adì 1603 19 settembre*, Stampata per Antonio Pinelli, Stampator Ducale, BNM. As early as 6 May 1380 a law authorized the Giustizia Veccchia, which was overseeing the forests of the Republic, to require *lignaroli* (wood haulers), before they entered the City, to secure "customs receipts and present them to the Justices within 24 hours, and avoid any hoarding or clandestine sale of wood" (Adolfo De Bérenger, *Saggio storico della legislazione veneta forestale dal secolo VII al XIX*, Venice, Libreria G. Ebhardt, 1863, p. 11). A law dated 6 March 1270 expressly prohibited the export of "unworked wood, or dried boards" [Latin] from Venice. See R[egia] Accademia dei Lincei, *Deliberazioni del Maggior Consiglio*, ed. R. Cessi, Bologna, Zanichelli, 1931, II, p. 31.

2. *Parte presa nell'Eccellentissimo Collegio delle Legne Adì 1603*, BNM. Similar regulation, with different measures, was imposed in the previous year on faggots coming from Mestre, the province of Treviso and Friuli: *Parte presa nell'Eccellentissimo Collegio delle Legne con intervento e ballottatione [sorteggio] delli Clarissimi Signori Sopraprovveditori e Provveditori alle Legne. 1602 Adì 2 settembre. Stampata per Antonio Pinelli Stampator Ducale*, BNM. The regulation that no wood was to be left to rot in the forests was broadly enforced as early as the 15th century. Cf. De Bérenger, *Saggio storico*, p. 16.

3. *Proclama dell'illustrissimo et Eccellentissimo Signor Gio. Battista Bon Sopra Provveditore, et Inquisitor nel Magistrato delle legne e Boschi*, ASV, PsB, b. l.

4. *Stampe varie dall'Anno 1595 al 1778*, ASV, PSB, b. 92. Document dated 26 November 1728.

5. Alvise Zorzi, *Una città una repubblica un impero. Venezia 967-1797*, Milan, Mondadori, 1984, pp. 82ff. On markers and minor structures, see Giovanni Battista Stefinlongo, *Pali e palificazioni della laguna di Venezia*, in *Lagunarie. Aspetti e caratteri della cultura materiale delle lagune venete e del territorio polesano*, eds. Mario Abis e Luciano Tedesco, Venice, Stamperia di Venezia, 1983, pp. 87ff.

6. M. F. Tiepolo, *Difese a mare*, in *Mostra storica*, p. 134. According to an undated petition, probably from the 17th century, there were 209 *"Pallade"* protecting the ports and the Lagoon, only 50 of which "are in a suitable condition" (*Suppliche e scritture al Senato*, ASV, SEA, b. 268).

7. *Scrittura del Magistrato alle Acque presentata al Senato Veneto il dì 18 Giugno 1740*, in *Documenti autentici*, pp. 5-6. The information given by Calcaneis is found in a document dated 30 December 1688, in *Scritture intorno alla laguna*, p. 2.

8. Cf. Raffaello Vergani, *Le materie prime*, in *Storia di Venezia*, XII, *Il mare*, eds. A. Tenenti & U. Tucci, Rome, Istituto dell'Enciclopedia Italiana, 1991, p. 287. For the Arsenale, which then covered 61,760 acres and employed 2,000 on the average, see Lane, *Storia di Venezia*, p. 418. For the "Heart of the Venetian State," see C. Gottardi, *Introduzione*, in U. Pizzarello e V. Fontana, *Pietre e legni dell'arsenale di Venezia*, Venice, Cooperativa Editoriale L'Altra Riva, 1983, p. 15. The Arsenale of course has a huge bibliography of its own: see the recent

contributions of Maurice Aymard and Ennio Concina in *Storia di Venezia*, XII, pp. 147ff.

9. Lane, *Storia di Venezia*, p. 418. The number of galleys traditionally held in reserve was 25; it was increased to 50 towards the end of the 15th century.

Chapter 2, Section 6 – Forest Management and Regrowth

1. De Bérenger, *Saggio storico*, p. 15; B. Vecchio, *Il bosco negli scrittori italiani del Settecento e dell'età napoleonica*, Turin, Einaudi, 1974, p. 55; I. Cacciavillani, *Le leggi veneziane sul territorio 1471-1789. Boschi, fiumi, irrigazioni*, preface by A. Zorzi, Limena, Signum, 1984, p. 90; and Vergani, *Le materie prime*, in *Storia di Venezia*, pp. 289-90. Private individuals occasionally bid on managing the forests for the State, as with the proposal by the owners of forests in the towns of Campolongo and Costa, "where no grass grows of any kind, and the sun can never penetrate, the chief reasons for harvesting the wood needed for the industry of the Arsenal" (*Scritture del Magistrato all'Eccellentissimo Senato, 1574-1581*, ASV, SEA, b. 270).

2. Cacciavillani, *Le leggi veneziane*, p. 90. Again and again during the following century laws were renewed emphasizing the prohibitions and reconfirming the regulations for the ten-year rotation of the *"legname dolce"* (understory) and cutting in State forests, and the seven-year rotation in town forests (De Bérenger, *Saggio storico*, p. 20).

3. Cacciavillani, *Le leggi veneziane*, p. 95. Besides de Bérenger's old but hard-to-find study, which underlines the non-bureaucratic nature of Venetian forest policy and its superiority as a precursor compared to the rest of the Italian and European States, we refer to the work of Cacciavillani for further and more detailed information on forest uses and legislation, on relations with local communities, transportation, etc. See also Vergani, *Le materie prime*. On the forests in the provinces of Treviso and Belluno, cf. Tentori, *Saggio sulla storia*, vol. XII, 1790, pp. 6ff. For a broader perspective, see also John Perlin, *A Forest Journey. The Role of Wood in Development of Civilization*, Cambridge (Mass.), Harvard University Press, 1991, pp. 43, 54, & 160.

4. *Tentori, Saggio sulla storia*, vol. VIII, p. 379. From 1590 on there were three Overseers for Montello. The prohibition regarding nobles was still in effect in Tentori's day. For Montello, where Venice ultimately betrayed her own policy of respecting local autonomy, see Cacciavillani, *Le leggi veneziane*, p. 100.

5. The quotations from Arsenal foreman Christofolo de Zorzi—who also recommended the phases of the moon as a guide for the periodic cutting—are in ASV, PsB, *Registro*, b. 150 bis, *Catastico d'Alpago*, 12 August 1638. On this forest, besides the study of Vergani, *Le materie prime*, p. 295, see the contribution of E. Zolli, in E. Casti Moreschi ed E. Zolli, *Boschi della Serenissima. Storia di un rapporto uomo-ambiente*, Venice, Arsenale, 1988, pp. 62ff., also useful for information on the rational cultivation of various species of trees.

6. *Scritture del Consiglio dal 1790 al 1797*, ASV PsB, b. 85.

7. *Stampe varie dall'Anno 1595 al 1778*, ASV PSB, b. 92. The chief recalled that in 1748, in those same valleys, an "incredible number of the same kind" of trees had been destroyed.

8. *Scritture del Consiglio dall'Anno 1790 al 1797*, ASV, PSB, b. 85.

Chapter 2, Section 7 – Forests and the Lagoon

1. *Proclama degli Illustrissimi e Eccellentissimi Signori Sopra Provveditori e Provveditori alle legne e boschi approvato il 29 settembre 1760,* in *Stampe varie dall'Anno 1595 al 1778*, ASV, PSB, b. 92.

2. *Deliberazioni del Maggior Consiglio di Venezia*, II, p. 319. These and later prohibitions in the 14th century concerned the "pine forests (*Pinete*) scattered here and there on the *lidi*, to avoid the danger of fire; it was prohibited to cut the same, and ordered not to uproot the canebrakes *(Canneti)*" in order not to weaken—as we have pointed out—the natural defenses against the sea (Tentori, *Della legislazione veneziana*, p. 55).

3. Sabbadino, *Discorsi*, p. 6. But see also Cacciavillani, *Le leggi veneziane.*

4. See *Un codice veneziano del 1600 per le acque e le foreste*, eds. Roberto Cessi & Annibale Alberti, Rome, Libreria dello Stato, 1934, p. 12.

5. *Stampe varie dall'anno 1595 al 1678*, ASV, PsB, b. 92.

Chapter 3, Section 1 – A Peculiar Political Arena

1. Cf. Cacciavillani, *Le leggi veneziane*, pp. 20ff. The institution of the *Collegio*, assisting the various tribunals, was in part a result of the same logic: see Zendrini, *Memorie storiche*, vol. II, p. 77. For a comparison with Dutch legislation, see Ciriacono, *Acque e agricoltura*, pp. 222ff. We also refer to this text for the development of hydrological science in the Lagoon, in constant and enlightening comparison with the case of Holland. For interesting comparisons of the two elites as to mentality and behavior, see Burke, *Venezia e Amsterdam.*

2. For all these aspects, see Tiepolo, *Difese a mare*, p. 134; Cessi, *Evoluzione storica*, p. 17; and Ugo Tomasicchio, *La conservazione dei Lidi a protezione della Laguna Veneta*, in *Laguna, fiumi, lidi, cinque secoli di gestione delle acque nelle Venezie*, Ministero dei Lavori pubblici—Magistrato alle Acque di Venezia (seminar held in Venice, 10-12 June 1983), Venice, Grafiche «La Press», n.d., II, 32, pp. 2ff.

3. Cucchini, *La laguna di Venezia*, p. 9.

4. *Suppliche e scritture al Senato* (undated documents), b. 268. To get an idea of the relative value of the expenditure, we note that the annual salary of an engineer in the 15th century was about 150 ducats.

5. *Documenti autentici*, p. 13. Cf. also, Tiepolo, *Difese a mare*, pp. 136-8; and Nice Antonia Benigni, *Gli interventi dei veneziani sui litorali veneti nel XVII sec.—I Murazzi*, in *Laguna, fiumi*, I-2 , pp. 5ff. For the following information as well, see Malaspina, *Memoria storico-idraulica*, p. 26.

Chapter 3, Section 2 – An Authority for Water

1. Zendrini, *Memorie storiche*, vol. I, pp. 9-10.

2. Cessi, *Evoluzione storica*, p. 19. In the course of the 14th century the Council of Ten was in charge of the Lagoon and sometimes the Senate, entrusting to the College of Sages the enforcement of decrees. Cf. Tentori, *Saggio sulla storia civile*, vol. VIII (1787), pp. 222ff.

3. Zendrini, *Memorie storiche*, I, p. 136. Particularly harsh penalties were provided for

whoever cut the levees along the Brenta for agricultural purposes. They were to have "their right hand cut off, one eye put out, and their possessions confiscated" (ibid., p. 136). On the structure of the Tribunal for Water, see also Cacciavillani, *Le Leggi veneziane*, p. 190.

4. Calcaneis, *Scritture intorno alla laguna*, p. 52. Periodic soundings were also made in the Lagoon, as provided by a decree dated 17 June 1690. Cf. Rompiasio, *Scrittura nell'universale*, p. 14.

5. Tentori, *Saggio sulla storia,* vol. VIII, pp. 223-4.

6. Ibid., p. 226. See also Ciriacono, *Acque e agricoltura*, p. 168. For the 5% tax, see *Metodo in pratica di sommario o sia compilazione delle leggi terminazioni e ordini appartenenti agl'illustrissimi e eccellentissimi Collegio e Magistrato alle Acque. Opera dell'Avvocato Fiscale Giulio Rompiasio*, Venice, 1733, critical edition by G. Carnato, Padua & Venice, Industria Grafica, 1988, p 528.

7. *Capitolare* (1530-1538), ASV, SEA, b. 43. In this document, the Sages establish a register on parchment (*carta membrana*) of information received from fishermen, to "better understand the course and movements of said Lagoon," and encourage the election by the fishermen's *Scuola* of San Nicolò of eight experts, who were to report periodically "under sacred obligation." Tentori also mentions the 1536 law: *Della legislazione veneziana*, p. 99. On fishermen as "experts," see also Ligabue, Basaglia, & Rossi Osmida, *Pesca e caccia*, p 159.

8. *Documenti Giustizia Vecchia*, ASV, b. C.

9. Cf. Zug Tucci, *Pesca e caccia in laguna*, p. 495. Sabbadino, *Discorsi sopra la laguna*, p. 45.

10. Rompiasio, *Scrittura nell'universale*, p. 8. On earlier destruction, see Tentori, *Della legislazione veneziana*, p. 178. Another negative opinion in Zendrini, in *Relazioni periti area la Laguna* (1725-1734), ASV, SEA, report dated 18 April 1726. The owners of fish pens were of course compensated "because (as the Sages wrote in 1661) a constant principle of the Republic is not to deprive subjects of their property without just cause," or, as was reconfirmed in 1725, "because the Senate's charity, even when confronted with its greatest interest, cannot leave without echo the arguments of private individuals who justly deserved compensation" (cf. A. Bullo, *La questione lagunare studiata sotto l'aspetto storico ed economico*, Florence & Rome, Tipografia Fratelli Bencini, 1884, pp. 9-10, who uses these documents to defend the controversial thesis that the Lagoon is not public domain).

Chapter 3, Section 3 – The Rule of Law

1. Cf. Sergio Escobar, *Il controllo delle acque: problemi tecnici e interessi economici*, in *Storia d'Italia*, Annali 3, *Scienza e tecnica nella cultura e nella società dal Rinascimento a oggi*, ed. G. Micheli, Turin, Einaudi, 1980, pp. 85ff. For the development of agriculture, in addition to Daniele Beltrami's old but detailed *Saggio di storia dell'agricoltura nella Repubblica di Venezia durante l'età moderna*, Venice & Rome, Istituto per la collaborazione culturale, 1955, pp. 9ff., see Angelo Ventura, *Considerazioni sull'agricoltura veneta e sull'accumulazione originaria del capitale nei secoli XVI e XVII*, in *Agricoltura e sviluppo del capitalismo*, Rome, Editori Riuniti, 1970, pp. 519ff., and Ciriacono, *Acque e agricoltura*, pp. 50ff.

2. Quoted in Ugo Mozzi, *I magistrati veneti alle acque ed alle bonifiche*, Bologna, Zanichelli, 1926, p. 25.

3. Ibid., p. 21.

4. *Relazioni dei periti 1557-1558*, ASV, PBI, b. 262, *Retratto della Brancaglia in Moncelese del 22 settembre 1558.*

5. Ibid., document dated 5 April 1582.

6. *Processi investiture ed altri atti*, ASV PBI, b. 427. Petition dated 6 March 1577.

7. Ibid., b. 262, document dated 8 March 1570.

8. Document dated 26 January 1530, in Cacciavillani, *Le leggi veneziane*, p. 138. See also *Un codice veneziano del 1600 per le acque e le foreste.*

9. *Parte presa nell'Eccellentissimo Collegio delle Acque 1579 adì 2 aprile In materia della Laguna. Stampata, per Antonio Pinelli Stampator Ducale a S. Maria Formosa, in Calle del Mondo Novo*, BNM.

10. Loc. cit.. On 16 January 1530 it was ruled that all private citizens must report to the *Collegio sopra le Acque* all forest lands cleared in the previous 40 years, on pain of confiscation for failure to report, and that 8% of cleared lands must be reforested within 11 months (De Bérenger, *Saggio storico della legislazione*, p. 21). Similar provisions may be found in the original version in Cacciavillani, *Le leggi veneziane*, pp. 138ff.

11. *Ordeni et Capitoli presi nel Collegio delli Clarissimi Signori Proveditori e Giustizieri Vecchi in Materia di ogni sorte di Rede, Spesse, Chiusse, Ostreghere e altro. Stampati infine 1599*, BNM.

12. Cacciavillani, *Le leggi veneziane*, p. 102.

13. *Suppliche e scritture sopra le acque* (1474-1580), ASV, SEA, b. 116.

14. Petition dated 4 August 1544, ASV, SEA, AZ, b. 20.

Chapter 3, Section 4 – Legality, Equality, and Liberty

1. K. A. Wittfogel, *Il dispotismo orientale*, Florence, Vallecchi, 1968. On the open, non-despotic character of community water use in a society that was not uniformly agricultural, cf. Piero Bevilacqua, *Le rivoluzioni dell'acqua. Irrigazione e trasformazioni dell'agricoltura tra Sette e Novecento*, in *Storia dell'agricoltura italiana in età contemporanea*, ed. P. Bevilacqua, I, *Spazi e paesaggi*, Venice, Marsilio, 1989, p. 269.

2. Cf. Silvano Avanzi, *Il regime giuridico della laguna di Venezia. Dalla storia all'attualità.* Venice, Istituto Veneto di Scienze, Lettere e Arti, 1993, p. 68.

3. Decision dated 18 February 1781, ASV, Stampe Giustizia Vecchia, b. C. At issue was the protection of young fry.

4. For all these aspects see Lane, *Storia di Venezia*, pp. 132ff. and 295-7. In 1719 there were 1703 Venetian noblemen over the age of 25 (the age for entering the Great Council), in a population of 140,000 (Burke, *Venezia e Amsterdam*, p. 26). On the restriction of earlier civil liberties, see Angelo Ventura, *Nobiltà e popolo nella società veneta del '400 e del '500*, Bari, Laterza, 1964. Cozzi has underlined the concentration of political power in the 16th century in the Council of Ten, in *Repubblica di Venezia*, pp. XIIff., and *Stati italiani: Politica e giustizia dal secolo XVI al secolo XVII*, Turin, Einaudi, 1982, pp. XIIff. On the effects of the *Serrata* of the Great Council, which hastened the decline and later led to abolition of the *concio*, or assembly of all freemen, cf. Giuseppe Maraini, *La Costituzione di Venezia dopo la serrata del Maggior Consiglio* (1931), facsimile edition Florence, La Nuova Italia, 1974, pp. 83ff.

5. See the complicated procedure, devised in order to defeat any possible kind of

alliance among group and family factions, followed in 1268 for the election of the Doge, in Lane, *Storia di Venezia*, p. 131.

6. Cf. Cozzi, *Repubblica di Venezia*, p. 102.

7. Lane, *Storia di Venezia*, pp. 102ff. A historian of the City, writing between the World Wars and reflecting the cultural climate of the time, referred to the "ideological superiority" of Venice—a significant term for us (Maranini, *La Costituzione di Venezia*, pp. 18-19). For the long history of the Venetian "myth," see Muir, *Il rituale civico*, pp. 17ff., and now also Crouzet-Pavan, *«Sopra le acque salse»*, II, pp. 970ff. For a longer historical perspective, with particular attention to the role played by art, see Eduard Hüttinger, *Il «mito» di Venezia*, in *Venezia—Vienna*, ed. Giandomenico Romanelli, Milan, Electa, 1983, pp. 187ff.

8. Lane, *Storia di Venezia*, p. 318. Maranini emphasized the fact that Venice was a "class-based regime" of the Venetian aristocracy, but "not a single provision of its constitution openly revealed that character" (*La Costituzione di Venezia*, II, p. 29).

9. Tentori, *Saggio sulla storia*, vol. II, p. 72. The author recalls that "In the year 1229 in the Council called *Pregadi* [the Senate], the political and economic interests of the Nation were committed to her trade."

Chapter 4, Section 1 – Daily Maintenance

1. Cornaro, *Della laguna*, p. 144.

2. Tentori, *Della legislazione veneziana*, p. 84.

3. "If I wished to relate the history of all the excavations made in the Lagoon and the Channels of the Estuary," Tentori writes, "I should have to compose a thick volume" (ibid., p. 198).

4. *Relazione de' Periti e Disposizioni*, 1583-1592, ASV, SEA, b. 158. Document by Alvise Zorzi dated 3 October 1589.

5. Quoted in *Laguna, lidi, fiumi*, p. 48. It must also be recognized, as noted in a document from the Sages dated 12 October 1531, that *"cavazione"* was also necessary for "the health and preservation of our city" (*Capitolare*, 1530-1538, ASV, SEA, b. 343). Again in 1736, in a report dated 17 October, Zendrini was passionate about a radical reëxamination of the problem of silting, in view of the great expenses required by constant dredging (*Relazioni dei periti ai Savi ed esecutori alle Acque circa la laguna*, 1735-1771, ASV, SEA, b. 56).

6. Tentori, *Della legislazione veneziana*, p. 86. The overseers were required to deliver to the Chief the *"Boletti di scarico,"* or summary of the material dredged: cf. Rompiasio, *Metodo in pratica di sommario o sia compilazione delle leggi, terminazioni e ordini*, pp. 18 and 73-5.

7. Avanzi, *Il regime giuridico della laguna*, p. 68. Non-complying owners were expropriated for an indemnity equal to the original value of the property. The quotation is from Filiasi, *Osservazioni sulle cause*, p. 98. For the process of gaining new ground in the Middle Ages by filling and draining, see Crouzet-Pavan, *«Sopra le acque salse»*, I, pp. 71ff.

8. In the 16th century dredged material was dumped in certain places that were specified periodically. General excavations in the Lagoon were carried out in 1546, 1658, and 1677, the last known as the *"escavazione generale."* Of course dredging in limited areas continued, like the *"scorticamento"* of the mud flats, etc. For all this, see Tentori, *Della*

legislazione veneziana, pp. 201-12.

9. ASV, SEA, AZ, b. 20.

10. Zendrini, *Memorie storiche*, vol. I. p. 81.

11. B. & L. Lanfranchi, *La Laguna dal secolo VI al XIV*, in *Mostra storica*, pp. 82-3. For greater detail, see now Crouzet-Pavan, *«Sopra le acque salse»*, I, pp. 89ff., and for a more juridical perspective, see Avanzi, *Il regime giuridico della laguna*, pp. 51ff.

12. The decisions quoted are all contained in Cacciavillani, *Le leggi veneziane*, pp. 192-7. For the Brenta and Brenta Nova, in the 17th century the various prohibions were still being repeated, always providing for restoration of the previous conditions at the expense of the guilty. Cfr. BNM, *1615 Adì 29 Ottobre. In materia delli Terreni de i Casamenti fatti della Brenta e del distrugger li ponti et li pascoli alli animali, e altro. Stampata per Antonio Pinelli Stampator Ducale.*

13. The reference is Ciriacono, *Acque e agricoltura*, pp. 50 and 69.

Chapter 4, Section 2 – Gigantic Undertakings

1. Cessi, *Evoluzione storica*, p. 23. For Marco Cornaro, see also G. Pavanello's detailed *Prefazione* to *Marco Cornaro (1412-1464), Scritture sulla laguna.*

2. Daniele Mainardi, *Stato presente della Brenta, suoi rovinosi effetti da Bassano all'ingiù [...]*, n. p. p., 1789, p. V.

3. Originally the Brenta did not empty into the Lagoon; the Paduans made the change in 1142, to protect their lands from flooding. It appears that the first writer to make this assertion, confirmed by subsequent historical and archeological studies, was Marco Cornaro, in *Scritture sulla laguna*, p. 8. For projects on the river see Cessi, *Evoluzione storica*, pp. 22ff.; Guido Ruggieri, *Alcuni significativi interventi sul Brenta*, in *Mostra storica*, pp. 121-3; and with a wealth of technical detail, Miozzi, *Venezia nei secoli*, in *La laguna*, pp. 97ff.

4. For a summary of the opinions of the various hydrologists see Camillo Vacani, *Della Laguna di Venezia e delle attigue provincie. Memoria*, Florence, Litografia degli Ingegneri, 1867, pp. 327ff.; Gian Albino Ravalli Modani, *Scrittori tecnici di problemi lagunari*, in *Mostra storica*, pp. 169ff.; and Ciriacono, *Acque e agricoltura*, p. 138. For the following information on changing the course of rivers, see the wealth of technical detail in Miozzi, *Venezia nei secoli*, in *La laguna*, pp. 104ff.

5. The boundary also signaled the beginning of a stricter and more detailed policy of prohibitions and obligations concerning the Lagoon, expressed over time in an abundant series of legislative actions. The demarcation by means of 99 stone markers was completed in 1792. For all this, see Avanzi, *Il regime giuridico*, pp. 78ff. For the territorial aspects of the demarcation project, see *I cento cippi di conterminazione lagunare*, ed. E. Armani, G. Carnato e R. Giannola, Venice, Istituto Veneto di Scienze Lettere e Arti, 1991.

6. The interests of the City and the Lagoon had a profound impact on the neighboring mainland economies. If the problem of silting in the Lagoon seemed to have been essentially solved, the same could not be said of the river systems on the mainland, which in some respects deteriorated over a long period of time. See Miozzi, *Venezia nei secoli*, in, *La laguna*, p. 111; Ciriacono, *Acque e agricoltura*, pp. 168-9. For the

oldest writings concerning the "rivers bordering the Venetian Lagoon," see Tentori, *Della legislazione veneziana*, pp. 23ff., and Filiasi, *Memorie storiche de' Veneti*, vol. VI, p. 122.

7. Francesco Marzolo, *Principali lavori eseguiti nella laguna di Venezia nel secolo XIX*, in *Mostra storica*, p. 220.

8. The record quoted above is one of a series of expert appraisals of the effects of the Piave in the course of the second half of the 17th century, in *Relazioni periti circa laguna e fiumi (1686-1689),* ASV, SEA, b. 139. The 19th-century judgment of the work is by Pietro Paleocapa, *Dello stato antico delle vicende e della condizione attuale degli estuari veneti*, Venice, Antonelli, 1876, p. 21. The author estimated the expenditure at 800,000 ducats, or $1,200,000. For greater detail on this and later information, see Miozzi, *Venezia nei secoli*, in *La laguna*, pp. 122ff., and Giorgio Tamba, *Alcuni significativi interventi sul Piave*, in *Mostra storica*, pp.125-7.

9. Miozzi, *Venezia nei secoli*, in *La laguna*, pp. 122ff.

10. Cf. Cucchini, *La laguna di Venezia e i suoi porti*, p. 14. For the Ficarolo break and its impact on Ferrara and its region, see Bevilacqua & Rossi-Doria, *Le bonifiche in Italia*, pp. 12-17.

11. *Relazione dell'avvocato fiscale Filippo de Zorzi del 9 gennaio 1595,* ASV, SEA, AZ, b. 20.

12. *Scritture circa le valli (1661-1666)*, ASV SEA, b. 127. For these issues see also Tamba, *Il taglio del Po a porto Viro* (1598-1604), in *Mostra storica*, pp. 129ff.

13. *Relazioni ed itinerari de' pubblici Periti 1503-1670*, ASV, SEA, b. 159, report dated 14 August 1611. In some locations, however, the expert recognized some visible signs of improvement.

14. Cucchini, *La laguna di Venezia e i suoi porti*, p. 15.

Chapter 4, Section 3 – Sunset for the Republic

1. Quoted in Cacciavillani, *Le leggi veneziane*, p. 99. Extensive 17th-century documents on the Adige are found in *Cavamenti nel Canal Grande e rii*, 1556-1557, ASV, SEA, b. 992.

2. See Giandomenico Romanelli, *Venezia Ottocento. L'architettura. L'urbanistica*, Venice, Albrizzi, 1988, p. 135.

3. See Miozzi, *Venezia nei secoli*, III, *La laguna*, p. 176. Bondesan, *L'evoluzione geologica*, in *Mostra storica*, pp. 36ff., and the biblography for the following notes.

4. For legislative aspects to date see Avanzi, *Il regime giuridico*, pp. 103ff.

5. The notice, issued on 22 June 1797 by the Health Commission of Venice, further ordered "*Capi Contrada* and *Facchini dei Campi* to see that the small basins at the foot of public wells are kept clean and filled with water" so that dogs may drink, and to lessen the risk of spreading rabies (*Raccolta di tutte le carte pubbliche stampate, ed esposte ne' luoghi più frequentati della città di Venezia*, Venice, Francesco Andrelo, 1797, vol. II, p. 277).

Chapter 5, Section 1 – The Revenge of the Land

1. For all these issues a very detailed reconstruction is given by Cucchini, *La laguna di Venezia e i suoi porti*, pp. 28ff. See also Maurizio Reberschak, *L'economia*, in Emilio Franzina, *Venezia*, Bari & Rome, Laterza, 1986, p. 232.

2. Alvise Zorzi, *Venezia austriaca. 1786-1866*, Rome & Bari, Laterza, 1985, p. 32. On the myth of Venice's decadence after the fall of the Republic, see Hüttinger, *Il "mito" di Venezia*, pp. 190ff.

3. Alvise Zorzi, *Venezia scomparsa. Storia di una secolare degradazione*, Milan, Electa, 1977, I, p. 177. See also Luigi Scano, *Venezia: terra e acqua*, afterword by E. Salzano, Rome, Edizioni delle Autonomie, 1985, pp. 26ff., who recognizes some positive aspects of the French administration, like compiling a new tax map and a new survey of the area surrounding the Lagoon.

4. According to an anonymous official document from the early 1830s: *Notizie statistiche intorno ai fiumi, canali, laguna e porti delle provincie comprese nel governo di Venezia*, Milan, Imperial Regia Stamperia, 1832, p. 49.

5. A student of Venice's urban context records the restoration of the tribunals early in the Austrian dominion. See Romanelli, *Venezia Ottocento*, p. 29. Of course, institutions have "a life of their own" and are not always brought to life by legislative action.

6. Gianpietro Zucchetta, *Un'altra Venezia. Immagini e storia degli antichi canali scomparsi*, Venice, Unesco-Erizzo Editrice, 1995, p. 55. For the number of waterways and the practice of filling, which in the 19th century alone was particularly widespread, see in addition Zucchetta, *Una fognatura per Venezia*, pp. 21ff. & 32.

7. Avanzi, *Il regime giuridico della laguna*, pp. 102ff.

8. By 1821 the population of the historic City had declined to 100,000. Cf. Romanelli, *Venezia Ottocento* , pp. 133ff.

9. See P. Paleocapa, introduction to Camillo Vacali, *Della laguna di Venezia e dei fiumi delle attigue provincie. Memoria*, Florence, Tipografia e Litografia degli Ingegneri, 1867, p. 3, relating how the danger was finally eliminated, and the role played by Vittorio Fossombroni, then in his eighties.

10. For these aspects, as well as the transformation of central Venice by the new territorial policy, see Cesco Chinello, *Porto Marghera 1902-1926. Alle origini del «problema di Venezia»*, preface by Silvio Lanaro, Venice, Marsilio, 1979, pp. 15-6. On the new economic prospects from the building of the railway, see Giovanni Distefano e Giannantonio Paladini, *Storia di Venezia, 1797-1997. La Dominante dominata*, Venice, Supernova Grafiche Biesse, 1996 pp. 283ff. Also Giannantonio Paladini, *Alla ricerca della conchiglia istituzionale perduta. Note per un profilo di Venezia nell'Ottocento,* in *Venezia. Itinerari per la storia della città*, eds. S. Gasparri, G. Levi & P. Moro, Bologna, Il Mulino, 1997, p. 358.

11. Avanzi, *Il regime giuridico della laguna*, p. 102. Various examples can of course be found in the holdings of the *Beni inculti*. See for example *Relazioni dei Periti 1557-1558*, ASV, PBI, b. 262.

12. A. Contin, *Del risanamento e della bonificazione dei bassifondi dell'estuario veneto in armonia colla conservazione lagunare. Considerazioni e proposte*, Venice, Comizio agrario e di pescicultura di Venezia, 1882, pp. 3 & 12.

13. Ibid., p. 8.

14. On these reclamations, mainly water projects, cf. Costante Bortolotto, *Bonifica ed agricoltura veneziane*, Venice, Tipografia del «Gazzettino Illustrato», 1931; Guido Emarcora, *Di alcuni lavori di bonifica in corso di esecuzione nel Veneto*, Venice, Officine Grafiche C. Ferrari, 1922; Vittorio Ronchi, *Bonifica di Cava Zuccherina*, Rome, Inea & Treves, 1930. For the

impetus for reclamation coming in that phase from private interests, see Emilio Franzina, *L'unificazione*, in *Venezia*, p. 42.

15. Avanzi, *Il regime giuridico della laguna*, pp. 143-4. See also Scano, *Venezia: terra e acqua*, p. 65.

Chapter 5, Section 2 – Rural Wealth and Urban Poverty

1. Scano, *Venezia: terra e acqua*, p. 34. Rolf Petri, *La frontiera industriale. Territorio, grande industria e leggi speciali prima della Cassa per il Mezzogiorno,* Milan, Franco Angeli, 1990, p. 57.

2. Petri, *La frontiera industriale*, p. 60, also for the use made of special legislation by the groups that founded Porto Marghera. See also Chinello, *Porto Marghera*, pp. 138ff.; Reberschak, *L'economia*, pp. 252ff., and Reberschak, *L'industrializzazione di Venezia (1866-1918)*, in *Venezia. Itinerari per la storia della città*, pp. 380ff.

3. Antonio Fradeletto, *Venezia antica e nuova*. Turin, Sten, 1921, pp. 248-9.

4. Scano, *Venezia: terra e acqua*, p. 60.

5. Giampaolo Rallo, *Guida alla natura nella laguna di Venezia*, Padua, Franco Muzzio, 1996, p. 14. See also, in general, Nico Ventura ed., *Le trasformazioni territoriali nell'area nord-orientale di Venezia*, preface by F. Indovina, Padua & Venice, Marsilio, 1975.

6. Nevertheless, it is significant that the population of the municipality grew from 317,000 (1951) to 346,000 (1981); starting in the teens it annexed to the old center new territories on the islands and the mainland (Mestre and Marghera). Cfr. Wladimiro Dorigo, *Venezia e il Veneto*, in *Storia d'Italia. Le regioni dall'unità a oggi. II Veneto*, ed. Silvio Lanaro, Turin, Einaudi, 1984, pp. 1050 & 1056.

7. Cfr Leonardo Ciacci e Giovanni Ferracuti, *Abitare a Venezia negli anni '80*, introduction by E. Salsano, Milan, Giuffrè, 1980, p. 13. This is a study made by the CRESME [Centro Ricerche Economiche Sociali di Mercato per l'Edilizia e il Territorio] for the Municipality of Venice that gives a less pessimistic view of signs of urban exodus.

8. In the 1931 census, for over 160,000 residents there were 28,000 dwellings, 12% of which were on the ground floor and frequently flooded: see Scano, *Venezia: terra e acqua*, pp. 62-3.

9. Report of the Minister for Public Works, Giacomo Mancini, 11 March 1969, in *La salvaguardia di Venezia nei dibattiti del Senato della Repubblica nella IV e V Legislatura*, Venice, Tipografia commerciale, 1980, p. 175.

10. Cf. Paolo Costa, *Venezia. Economia e analisi urbana*, Milan, Etas libri, 1993, pp. 136ff. See also Ventura, *Le trasformazioni territoriali nell'area nord-orientale di Venezia*, p. 800.

11. In 1981, 27,000 workers came into Venice daily from the province alone, while only 2,000 left Venice (Reberschak, *L'economia*, p. 272).

Chapter 5, Section 3 – Political Decay and Industrial Giants

1. Rallo, *Guida alla natura*, p. 66. See also Gianni Caniglia, *Il paesaggio vegetale*, Michele Neugebauer, Francesco Scarton, Massimo Semenzato, *L'avifauna lagunare*, and Giampaolo Rallo, *I mammiferi*, all in *Laguna. Conservazione di un ecosistema.*

2. However, timely warnings of danger were given: see Armando Scarpa, *Si vuol*

distruggere Venezia e la sua laguna?, Venice, 1962, pp. 13ff. By the middle of the 70s statistics showed the increase in oil-tanker traffic and processing of petroleum products. Cf. Marino Folin, *Lo sviluppo del porto di Venezia*, in *Venezia: prospettive di sviluppo e politiche di governo*, ed. Donatella Calabi, Venice, Marsilio, 1976, pp. 124-5.

3. R. Cossu, E. De Fraja Frangipane, *Stato delle conoscenze sull'inquinamento della laguna di Venezia*, Ministero dei Lavori pubblici, Magistrato alle Acque, Venice, Consorzio Venezia Nuova, (n.d. [late 80s]), p. 63.

4. See the Report to the Senate, dated 19 May 1964, by the Minister for Public Works, Pieraccini, recognizing the Mayor of Venice's opposition to oil tankers sailing past St. Mark's, and announcing the forthcoming opening of the Malamocco Channel, in Comune di Venezia, *La salvaguardia della laguna nei dibattiti*, p. 64.

5. Alberto Scotti, *Progettazione delle opere di difesa dalle acque alte*, in «Quaderni trimestrali del Consorzio Venezia Nuova», 1993, 3.

6. Cf. *Venezia 1966-1996. 30 anni di salvaguardia raccontati attraverso la stampa*, prefazione di M. Cacciari, Rome, ANSA [news agency], 1996, p. 149.

7. Testimony of Roberto Ferrigno, representing Greenpeace Italia (and other environmental organizations: WWF, Legambiente, Estuario nostro), 10 May 1991, in Atti parlamentari, Camera dei Deputati, X legislatura, Indagini conoscitive e documentazioni legislative. Commissione VIII (Ambiente, territorio, lavori pubblici), Comitato permanente per i problemi di Venezia, *Audizioni relative all'indagine conoscitiva n. 55 sullo stato di attuazione della, legislazione speciale per la salvaguardia di Venezia*, Rome, 1992, pp. 128-9.

8. Cossu & De Fraja Frangipane, *Stato delle conoscenze sull'inquinamento della laguna*, p. 67; Roberto Stevanato, *L'inquinamento atmosferico nell'area veneziana*, in *A vent'anni dall'evento di marea*, pp. 153ff.

9. Consorzio Venezia Nuova, *L'inquinamento di origine agricola nella laguna di Venezia*, Venice, 1989. In this connection see also *Relazione sullo stato dell'ambiente* (1989) by the Ministry for the Environment: Atti parlamentari, Camera dei Deputati, Servizio studi, *Documentazione per le commissioni parlamentari. Interventi per la salvaguardia di Venezia. Documentazione di base.* Roma n.d. [1990], especially p. 254 for the increased presence of nutrients in the Lagoon.

10. Cf. Zucchetta, *Una fognatura per Venezia*, pp. 187ff., also emphasizing the damage caused by wave action from vessels to private drains, which discharge matter in the vicinity and erode the fabric of buildings.

11. Cf. *Le attività per la salvaguardia di Venezia e della sua laguna.* Supplement to «Quaderni trimestrali del Consorzio Venezia Nuova», 1997, 1. The system of sewers and treatment plants serving mainland Venice was initiated by the Region in 1979 and based on an old existing plan. See Consorzio Venezia Nuova, *Stato di attuazione del piano direttore per il disinquinamento della laguna di Venezia*, ed. Eugenio de Fraja Frangipane, n.p.d. [Venice 1988], pp. 22-3.

12. Cf. *Il consorzio Venezia Nuova dal 1987 al 1992*, in «Quaderni trimestrali del Consorzio Venezia Nuova», 1993, 1, and *L'inquinamento idrico della gronda lagunare*, ed. E. Fortuna, Padua, Cedam, 1988. For the physical and environmental problems facing the Lagoon today, see *Il recupero morfologico della laguna di Venezia*, Supplement to «Quaderni trimestrali del Consorzio Venezia Nuova», Venice, 1993.

13. For all this, see *Un guasto, petrolio in laguna a Venezia*, in «l'Unità», 2 December 1995 , and ANSA, op. cit., Venice, 1966-1996, p. 149.

14. For the plan to close the Marghera pipeline—which supplies the refineries in Mantua—substituting for it the Genoa-Cremona pipeline, see *Le attività per la salvaguardia di Venezia*, p. 36.

Chapter 6, Section 1 – As in the Beginning, Threats From the Sea

1. G. Cecconi, *Venezia e il problema delle acque alte. Il rischio di danno al patrimonio urbano a causa della crescita relativa del livello del mare*, in «Quaderni trimestrali del Consorzio Venezia Nuova», 1997, 2.

2. Giuseppe Creazza, *Aspetti del degrado strutturale a Venezia*, in Istituto Veneto di Scienze, Lettere e Arti, *A vent'anni dall'evento di marea del novembre 1966*, Atti della giornata di studio, Venezia 1987, p. 96.

3. Zorzi, *Venezia scomparsa*, p. 285.

4. Giulio Obici, *Venezia fino a quando?* Padua, Marsilio, 1967, p. 13. In the same text, see Cesare De Michelis, *Cronache dell'acqua alta*, pp. 43ff.; ANSA, Venice, 1966-1996, pp. 3ff.

5. Ibid., p. 23.

6. *Le attività per la salvaguardia*, p. 20.

7. *La laguna di Venezia e i suoi porti*, p. 45.

8. Ibid., p. 48.

9. Scotti, *Progettazione delle opere.*

10. Cf. Consorzio Venezia Nuova, *Il restauro della laguna. Le casse di colmata*, Venice, 1988; *La laguna di Venezia. Tendenze evolutive*, in *Il recupero morfologico della laguna*, p. 7.

11. Paolo Rosa Salva, *I rapporti tra salvaguardia ambientale e attività produttive nella legge speciale, negli indirizzi e nel bando di concorso-appalto*, in Calabi, *Venezia: prospettive*, p. 134 (his italics).

12. The Lagoon is also suffering from the abandonment of the drainage area *(zona di gronda)* and the buildings in it, as well as from tampering with its old installations (traps *[sifoni]*, drains *[chiaviche]*, gates *[chiusure]*, etc.). See the testimony of Giampaolo Rallo, in *Audizioni relative all'indagine conoscitiva*, p. 132.

13. The process of subsidence is very old. Excavations made in 1885 at the foot of the Campanile of St. Mark's revealed an older foundation (888-912) that sank in the course of the centuries. Cf. Pietro Leonardo, *Dati e problemi idrogeologici, meteorologici e paleontologici veneziani*, in *A vent'anni dall'evento di marea*, p. 11.

14. For all the aspects briefly summarized here, now benefiting from a huge bibliography, see Unesco, *Rapporto su Venezia*, preface by R. Mahhen, Milan, Mondadori, 1969; Eugenio Miozzi, *Venezia nei secoli*, IV, *Il salvamento*, Venice, Il Libeccio, 1969; Comune di Venezia, *La salvaguardia fisica della laguna. La regolamentazione dei livelli marini in laguna e la difesa degli insediamenti urbani dalle «acque alte»*, Venice, Marsilio, 1983; Comune di Venezia-WWF, *Laguna. Conservazione di un ecosistema*; Scotti, *Progettazione delle opere di difesa.*

15. Eugenio Miozzi and Giuseppe Miozzi, *Come difendere Venezia dallo sprofondamento. Studio generale delle pressurizzazioni nel territorio influente nel bacino lagunare di San Marco alfine di evitare ulteriori sprofondamenti.* Venice, Il Libeccio, n.d. [1974], p. 5. The authors propose

halting withdrawals from the water table and refilling the wells with water, as in several successful experiments abroad.

16. Local subsidence as a result of human activity stopped in 1970, when withdrawals from the water-table ended. Overall, in the course of the century the sea level rose 11.3 cm, man-made subsidence till 1970 was 7.7 cm, and regional subsidence was 4 cm. Cf. Cecconi, *Venezia e il problema delle acque alte*, p. 25. See also Luciano Lippi, *Problemi ambientali di Venezia*, in *Venezia e i problemi dell'ambiente. Studio e impiego di modelli matematici*, presentation by G. Puppi, Bologna, Il Mulino, 1975, p. 24.

Chapter 6, Section 2 – Strategic Responses and Global Threats

1. For the debates in this phase, see Comune di Venezia, *La salvaguardia della Venezia;* Scano, *Venezia: terra e acqua*, pp. 138ff.; ANSA, *Venezia 1966-1996.*

2. Testimony by speakers and the political debate over the problems of Venice from the early 1980s to our decade are collected in Paolo Cacciari, *La salvaguardia della Venezia. Dieci anni di battaglie*, presented by S. Scaglione, Venice, ARC, 1995.

3. For a more current summary see *Studi, progetti, opere. Aprile 1996-marzo 1997*, in «Quaderni trimestrali del Consorzio Venezia Nuova», 1997, 1.

4. Program address by Mayor Massimo Cacciari, given on the occasion of municipal elections in November 1997. Unpublished. My thanks to Dr. Leopoldo Pietragnoli for making available to me the information produced by the Cacciari administration.

5. Edoardo Salzano, *Postfazione. La sfida di Venezia*, in Scano, *Venezia: terra e acqua*, p. 433.

Chapter 6, Section 3 – A Story Just Begun: Rescue

1. *Le attività per la salvaguardia della laguna.*

2. *Le casse di colmata*, Consorzio Venezia Nuova, Venice, 1988. The WWF has created the first protected area in the Lagoon, Valle Averto. See the testimony of Giampaolo Rallo, in *Audizioni relative all'indagine conoscitiva*, p. 132.

3. G. Cecconi and P. Nascimbeni, *Ricostruzione e naturalizzazione delle dune artificiali sul litorale di Cavallino*, in «Quaderni trimestrali del Consorzio Venezia Nuova», 1997, 2.

4. M. Gentilomo, *Considerazioni intorno alla difesa delle fasce costiere. I litorali della laguna di Venezia*, ibid., p. 63.

5. See Chapter I, note 5 above.

6. See M. Gentilomo, *Venezia. Il dibattito sulle opere di difesa dalle acque alte eccezionali*, in «Quaderni trimestrali del Consorzio Venezia Nuova», 1995, 4 & 1996, 1.

7. Even though protective works are planned for those places as well: see *Le attività per la salvaguardia della laguna*; *Interventi di difesa dalle acque alte nei centri urbani lagunari*; *Interventi di difesa dalle acque alte nel centro urbano di Malamocco*, as well as the historical contribution by Giorgio Bellavitis, *La pratica di «elevare» la quota delle pavimentazioni e le tradizioni degli abitati lagunari nella difesa locale dalle acque alte*, all in «Quaderni trimestrali del Consorzio Venezia Nuova», 1993, 2.

8. See *4* [i.e. *Quattro*] *anni di lavoro per il futuro di Venezia e Mestre*, Cacciari Administration,

n.p.d. (a collective document drawn up by the Municipal Council before the elections in November 1997).

9. *Stato delle conoscenze sull'inquinamento della laguna*, p. 71.

10. Cf. *Relazione del Presidente del Consorzio Venezia Nuova [Luigi Zanda] sulla gestione dell'esercizio 1994*, in «Quaderni trimestrali del Consorzio Venezia Nuova», 1994, 4 and 1995, 1. For Mo.S.E., see Ludovico Solinas, *Modulo sperimentale elettromeccanico—Mo.S.E. I risultati di quattro anni di sperimentazioni*, loc. cit. For all these aspects, see *Il Consorzio Venezia Nuova dal 1987 al 1992*, and Alberto Scotti, *Progettazione delle opere di difesa dalle acque alte*, II, in «Quaderni trimestrali del Consorzio Venezia Nuova», 1993, 4 & 1994, 1.

11. *4 anni di lavoro,* Cacciari Administration.

12. loc. cit.

13. Leonardo Benevolo, *Venezia non è condannata a morte*, in *Le grandi città italiane. Saggi geografici e urbanistici*, ed. R. Mainardi, Milan, Franco Angeli, 1971, p. 214.

14. See the testimony of G. Ruffolo, 6 February 1991, in *Audizioni relative all'indagine conoscitiva*, pp. 11-12.

15. E. Tiezzi and N. Marchettini, *Cambiamenti climatici e comparazione evolutiva degli ecosistemi antropizzati e degli ecosistemi naturali. Le implicazioni per la laguna di Venezia*, in «Quaderni trimestrali del Consorzio Venezia Nuova», 1997, 2.

Localities, Rivers, and Regions

Mentioned in *Venice and the Water.*
Those especially useful in following Bevilacqua's account are in bold.

Adige
Agordo
Ammiana
Bacchiglione
Bassano
Belluno
Bondante
Bottenigo
Brenta
Brondolo
Bruson (San Bruson di Dolo)
Ca' Deriva
Cadore
Carnia
Chievo
Cittanova Eracliana
Comacchio
Conche
Cortellazzo
Costanziaco
Dogaletto
Dolo
Equilohttp
Fiesolo
Fossone
Friuli
Giudecca
Livenza
Marghera
Misericordia
Montalbano
Montello
Monte San Lorenzo
Montiron
Moranzano
Motta dei Cunicci
Orbetello
Oriago
Pellestrina
Piave
Po
Porto Viro
Primaro
Sile
San Pietro in Volta
Santa Cristina
Sant'Adrian
Santa Margherita
Santa Maria, Lugo
Sant'Erasmo
Sottomarina
Tagliamento
Torre Astura
Treporti
Treviso
Tronchetto
Verona
Veronese de' Marzi
Volano